WITHDRAWN

The Legacy of
George Ellery Hale

The MIT Press
Cambridge, Massachusetts,
and London, England

The Legacy of
George Ellery Hale

Evolution of
Astronomy and
Scientific
Institutions,
in Pictures
and Documents

Edited by
Helen Wright,
Joan N. Warnow,
and
Charles Weiner

Copyright © 1972 by
The Massachusetts
Institute of Technology

This book was designed and produced by
The MIT Press Media Department.
It was set in Linofilm Trade Gothic
by Dynamic Composition, Inc.
printed on Oxford Sheerwhite Opaque
by Halliday Lithograph Corp.
and bound in Interlaken AL1 557-8 Matte
by Halliday Lithograph Corp.
in the United States of America.

All rights reserved. No part of this
book may be reproduced in any form
or by any means, electronic or
mechanical, including photocopying,
recording, or by any information
storage and retrieval system, without
permission in writing from the
publisher.

ISBN 0 262 23049 6 (hardcover)

Library of Congress catalog card
number: 74-148854

1
George Ellery Hale
1868–1938

Introduction
vi
Chicago Childhood
2
College Days at M.I.T.
9
Kenwood and Yerkes
18
Mount Wilson
42
National and International Affairs
72
Pasadena as a Cultural Center
87
The Hale Solar Laboratory and the 200-inch Telescope
97

2
Selected Papers of George Ellery Hale

Introduction
114
Photography of the Solar Prominences
Thesis for the B.S. in physics (M.I.T., 1890)
117
A Plea for the Imaginative Element in Technical Education
145
Solar Vortices and Magnetic Fields
155
National Academies and the Progress of Research
177
The Possibilities of Large Telescopes
193

3
Perspectives

Introduction
207
Astronomical Telescopes since 1890
C. Donald Shane
209
Astronomical Instrumentation in the Twentieth Century
Ira S. Bowen
239
Research on Solar Magnetic Fields from Hale to the Present
Robert Howard
257
Hale and the Role of a Central Scientific Institution
in the United States
Daniel J. Kevles
273
Index
289

Introduction

In 1889 a young undergraduate at the Massachusetts Institute of Technology invented a new instrument that could obtain valuable information never before available about the sun. When he died almost fifty years later, his inventions of instruments and institutions had transformed man's knowledge of the universe and the means of obtaining that knowledge. George Ellery Hale's legacy to the world, better known than his name, is symbolized in the now familiar dramatic photographs of solar flares, star clusters, and distant spiral galaxies. These reveal at once the growth in our knowledge, the enormity of the gaps in our understanding, and the complexity and power of the instrumentation that makes possible such explorations. They continue to provide new evidence of Hale's outstanding ability to conceive, plan, finance, and build new types of research institutions linking astronomical telescopes of unprecedented size with modern physics laboratories. In recognition of his role, the Mount Wilson and Palomar Observatories, which he founded, were renamed the Hale Observatories in 1970.

Hale's legacy also includes his concepts of the role of scientific societies and journals and of educational and cultural institutions in the sciences and humanities. *The Astrophysical Journal,* National Academy of Sciences, International Astronomical Union, International Council of Scientific Unions, Henry E. Huntington Library and Art Gallery, and the California Institute of Technology all bear the stamp of Hale's philosophy and organizational genius.

Another, more traditional, part of the Hale legacy is the rich personal record of how and why he worked to achieve these results. Preserved and available for scholarly study are his diaries, original manuscripts, and voluminous correspondence, which reveal his day-by-day activities. These documents, and related photographic materials, supplement his published writings and add immensely to our understanding of Hale's life, work, and influence. In this volume we present Hale's legacy through original documents and photographs, selections from his writings, and authoritative reviews of the major twentieth-century developments in fields upon which Hale had enormous impact.

In Part 1, the essay by Helen Wright, Hale's biographer, provides an overview of his life and work. The essay also serves as a narration for a sequence of related photographs, newspaper clippings, and original letters. Most of these materials were originally assembled by Helen Wright and Joan Nelson Warnow in the Niels Bohr Library of the American Institute of Physics for a joint American Association for the Advancement of Science and AIP exhibit at the 1968 annual meeting of the AAAS in connection with a symposium arranged by Charles Weiner, celebrating the hundredth anniversary of Hale's birth. Since then many of the materials have been displayed in the Niels

Bohr Library in New York, and the Library has assembled and distributed a touring version of the exhibit which has been displayed at twenty institutions in the United States and Canada.

Part 2 consists of five selections from Hale's writings showing his own scientific research (including his previously unpublished M.I.T. thesis), his efforts to restructure scientific education and the professional organizations of science to serve science and the public better, and his vision of large-scale astronomical telescopes and observatories. Part 3 includes essays surveying the development of large telescopes and related instrumentation, solar research, and scientific organizations — all relating Hale's original contributions to subsequent developments. These essays were originally presented by their authors at the AAAS Hale Centennial Symposium in 1968, which was chaired by Nicholas U. Mayall.

The editors acknowledge, with deep gratitude, the wholehearted cooperation and assistance in making documents available that have been given by Horace W. Babcock, director of the Hale Observatories, and the Observatories staff, especially William Miller, who provided many of the photographic materials. We also appreciate the support of the American Institute of Physics for sponsoring and housing our efforts in the Niels Bohr Library. The collections of the Library, made possible by support from the Friends of the Niels Bohr Library, have provided valuable resources for this book. The American Association for the Advancement of Science supplied the initial impetus through its support of the original exhibit and symposium. The AAAS meetings editor, Walter Berl, provided special encouragement. Our thanks also go to Margaret Hale Scherer, Mary Lea Shane, Judith Goodstein, Margaret Harwood, and Ronald Watson and the many other individuals, librarians, and archivists who have made source materials available. Indispensable aid in the preparation of the manuscript was provided by Dorothy Schreibersdorf of the Niels Bohr Library.

Helen Wright, Joan Nelson Warnow, Charles Weiner

October 1971

A preparatory drawing for the oil portrait of Hale by Seymour Thomas, which is hung in the National Academy of Sciences building in Washington, D.C.
The Hale Observatories.

George Ellery Hale
1868—1938

Helen Wright

Chicago Childhood

In 1932 George Ellery Hale received the Copley Medal, highest honor of the Royal Society of London. Soon he discovered he was to be further honored as a "Scientific Worthy" by the British journal *Nature*. While he appreciated this recognition of his achievements, he found it difficult to picture himself as a "true Worthy." Highly amused, he wrote to Harry Goodwin, his former classmate at the Massachusetts Institute of Technology, "If you know a good beard maker, please let me know at once. For what is a 'Worthy' without a long grey beard? Luckily my old friend [Hugh Newall] is to write the article, and he has a good sense of humor. For to put me in such a list, would draw a smile from the deepest pessimist."[1]

In response to a request for information from Newall, Hale recorded many of the details of his life in a way he might never have done otherwise. These "Biographical Notes" illuminate that life for us today. From them we gain a picture of his family background in Chicago, where he was born at 236 North La Salle Street, and of his boyhood on Drexel Boulevard in "the suburbs" to which the family had fortunately moved before the Great Fire. We learn of his father, William Hale, from whom he inherited the intense energy, the engineering and organizing ability that would drive him to great accomplishment in his chosen field, astrophysics. We learn of the influence of his mother on the development of his creative imagination through the books he read, ranging from *Grimm's Fairy Tales* to *Don Quixote*, from *The Iliad* and *The Odyssey* to the poetry of Keats and Shelley, from *Cassell's Book of Sports and Pastimes* to Jules Verne's *From the Earth to the Moon*.

We follow him from the Oakland Public School to the Allen Academy and the Chicago Manual Training School where he took shopwork. We discover his aversion to the "confinements and the fixed duties of school life":

Born a free lance, with a thirst for personal adventure, I preferred to work at tasks of my own selection. As a boy, largely through the constant encouragement of my father, I became interested in tools and machinery at a very early age; I always had a small shop with tools, first in the house, and later in a building of my own construction in the yard. I also had a little laboratory where I performed simple chemical experiments, made batteries and induction coils, worked with a microscope etc. After construction of a small telescope for myself, my father bought me an excellent 4-inch Clark. I used this constantly, but my enthusiasm reached the highest pitch when I learned something about the spectroscope. My greatest ambition was to photograph a spectrum and this I soon succeeded in doing with a small one prism spectroscope purchased for me by my father. I think this was in 1884. Solar spectroscopic work appealed to me above all things and I read everything I could find on the subject. My father always bought for me any books that I needed, but in the case of instruments his policy always was to induce me to construct my own first and then to give me a good instrument if my early experiments were successful. In 1888 he built for me, after my own designs, a spectroscopic laboratory in which a Rowland concave grating of 10 feet focal length was erected. This was the nucleus of the Kenwood Observatory.[2]

Hale entered astronomy at a time when the majority of astronomers were concerned with the positions, motions, and distances of the stars and evinced little interest in their physical nature. Pioneer work had been done in Italy, England, France, and Germany, but the science of astrophysics was still in an embryonic state.

Hale's influence on the development of astrophysics was immense. Perhaps no other single individual contributed more to the rise of modern astrophysics. His tools were those he had used in boyhood: the telescope, the spectroscope, the photographic plate; his working place the observatory, combined with laboratory and shop where astrophysical problems might be solved to discover the physical nature of the universe.

Hale as a young boy.
Niels Bohr Library.

Hale's "Biographical Notes," 1933. *The Hale Papers, in the California Institute of Technology Archives and, in part, in the Hale Observatories Library (hereafter cited as the Hale Papers).*

Drexel Boulevard, described in *Chicago and Environs* as "the most beautifully laid out boulevard in Chicago—...modeled after the Avenue de l'Impératrice in Paris."
Chicago and Environs (Chicago: Schick, 1891).

Interior of the house at 4545 Drexel Boulevard. The Hale family moved here in 1886.
The Hale Papers.

5   Chicago Childhood

Observations by George Hale of microscopic bodies found in the ditches of Kenwood, a Chicago suburb.
The Hale Papers.

List of Microscopic Objects.

No.	Name	Medium	Date
1.	Cyclops vulgaris	Glycerine	4/2/84
2.	Antenna of gnat	Dry	4/12/84
3.	Cyclops vulgaris	Glycerine	4/3/84
4.	Sponge	Dry	4/9/84
5.	Branchipus Stagnalis	Glycerine	4/2/84
6.	Eggs of Cyclops	Glycerine	4/2/84
7.	Branchipus Stagnalis muscle	Balsam	
8.	Muscle from frogs leg	Glycerine	
9.	P. fasicularis fungus	Dry	3/16/84
10.	Wing of Promethea Moth		4/16/84
11.	Leg of spider	Dry	
12.	Wing cover of – beetle	Dry	
13.	Wing " " "	Dry	
14.	Antenna " " "	Dry	
15.	Skin of frog	Dry	
16.	Wing of house-fly	Dry	4/12/84
17.	" " gnat	Dry	4/12/84
18.	Body " "	Dry	4/12/84
19.	Scales of – Moth	Dry	4/12/84
20.	Lung of frog	Nat. injection Gly.	4/8/84
21.	Lung of frog	Gly.	
22.	Wing of – fly	Dry	4/15/84
23.	Wings of –	Dry	4/15/84

No.	Name	Medium	Date
24.	Potato starch	Dry	4/18/84
25.	Scale of perch	Balsam	4/18/84
26.	" " "	Dry	4/18/84
27.	Fin " "	Dry	4/18/84

450   RECREATIVE SCIENCE.

may now be fitted to the side *b d*, and then your box is fit to take to the tinner again, to have a lid soldered on the top, with a small hole in it for filling the box with disulphide of carbon, and this hole must have a screw in it, so that the screw stopper may afterwards be made air-tight to prevent the evaporation of disulphide.

The slit (Fig. 71)..—Procure two pieces of brass plate, and cut into corresponding semicircles. Make their straight sides as smooth as possible, by filing with a very smooth file, and then wrapping the file in some fine emery cloth and rubbing perfectly smooth with this (*a*). Make another plate circular (*b*) with a square opening in the middle of the length of the required slit, and on this plate rivet the two brass semicircles, as in Fig. 71, *c*, so that the finest slit possible is seen when they are held up to the light. If, however, you prefer to have a slit of the ordinary type, you may, if you possess skill enough, attempt to make one with the sliding shutter arrangement, as at s, Fig. 62.

Fig. 71.—Making the Slit.

Fig. 72.—Collimating Tube.

Perhaps the simplest way of all is to take a circular plate of brass, of the same diameter as the tube it has to be fitted to, and rule a straight line along its diameter. This straight line is now graven into by means of the fine point of a file until it is nearly worn through, and now the slit may be cut through gently with the thin, sharp blade of a knife. It is such a slit as this which is employed in the spectroscope, Fig. 73.

Fig. 73.—Home-made Spectroscope.

The collimating tube (Fig. 72).—Procure a double convex lens of about a foot, or rather less, focus. It may be the glass out of an old pair of spectacles. A paper tube must now be made, of rather less than the focal length, and of the

Homemade spectroscope used as a model by young Hale.
Cassell's Book of Sports and Pastimes (London: Cassell, Peter, Galpin, 1881).

The astronomer, Sherburne W. Burnham, whom Hale met when he was about thirteen. "Through him I learned of a second-hand 4-inch Clark refractor.... I mounted it on the roof of the house. The astonishing views it afforded of Saturn, the Moon, Jupiter and other objects excited an intense desire to carry on actual research."
Photo by Dorothy Wallace, 1908. Sky and Telescope.

College Days at M.I.T.

In 1886 Hale entered the Massachusetts Institute of Technology. He majored in physics under Professor Charles Cross. Yet, in contrast to his own laboratory work, he found most of the courses dull and uninspiring. In his spare time he read and abstracted everything he could find on astronomy and spectroscopy at the Boston Public Library. He also persuaded Edward C. Pickering, director of the Harvard College Observatory, to let him work there as a volunteer assistant. While there, he continued to puzzle over a problem that had long interested him: to photograph the solar prominences at the sun's limb in full daylight so that a permanent record of these and other solar phenomena could be made. The solution came to the twenty-one year old one day during the summer vacation after his junior year. In August 1889, he was riding along on a Chicago trolley car when the idea came to him "out of the blue." He wrote to his classmate and close friend, Harry Goodwin, to tell him about it:

Of scientific work I have accomplished but one thing this summer, and even that did not involve much labor. It is a scheme for photographing the prominences, and after a good deal of thought I can see no reason why it will not work. The idea occurred to me when I was coming home from uptown the other day, and amounts to this — Stop the clock of the equatorial and let the sun transit across the slit, — which is radial to the limb. Bring h (in the blue) into the field of the observing telescope, and replace the eye-piece by a plate holder, held in a suitable frame, and drawn by clockwork across the field at the same rate as the sun crosses the slit. As the h line lengthens and shortens, — as it will do with the variable height of the prominences — the plate will photograph its varying lengths side by side, and thus produce an image of the prominence.

That is the idea in the rough, but I have studied it out in detail, and designed a travelling plate holder, which I will have [John] Brashear make. I have also got an arrangement by which all fog is avoided, and I have great hopes that the thing will be a success. If it is, new chances for work on the prominences will be opened, as in this way the changes during short intervals of time can be noted with much greater accuracy than in drawings. . . . Let me know what you think of the plan.

I am not going to say anything to Chas. [Cross] or even Prof. Pickering about it until I find out whether it will work, but I am going to write Prof. Young to-day, so as to have a record of it in case it should amount to anything.[3]

This was the instrument Hale was to call a spectroheliograph. In the fall of 1889 he tried out his principle at the Harvard College Observatory, and in his M.I.T. senior thesis, "Photography of the Solar Prominences" (see page 130) he described the results that proved the feasibility of the method. Out of these beginnings was born his lifelong interest in that "typical star," our sun, the only star close enough to study in detail.

"The Massachusetts Institute of Technology, when I entered it in 1886, was the leading scientific and engineering school in this country. I selected the course in Physics, which was also taken by Harry M. Goodwin, who became my closest friend." The Institute was then in the heart of Boston, at the corner of Boylston and Clarendon Streets on Copley Square.
Massachusetts Institute of Technology Catalogue. Boston: 1871.

George Hale
Photo by Notman.

Harry M. Goodwin, Hale's classmate, and Arthur A. Noyes, his instructor in chemistry and his lifelong friend.
Photos by Notman.

Students at M.I.T., about 1890.
The Hale Papers.

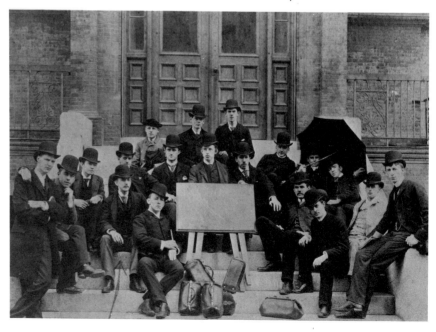

10 George Ellery Hale 1868-1938

Transcript of Hale's college record
1886-1890.
Explanation of Grades
H: passed with honor
C: passed with credit
P: passed
Massachusetts Institute of Technology.

Harvard College Observatory. *Daniel W. Baker, History of the Harvard College Observatory During the Period 1840-1890 (Cambridge: 1890).*

Hale to Edward C. Pickering, director of the Harvard College Observatory, February 26, 1888, asking permission to work there on Saturdays. *Harvard University Archives.*

Charles Young, the leading American authority on solar physics, whom Hale met at Princeton in 1889. "My first view of a solar prominence was obtained with Professor Young's telescope, and I began to search for a satisfactory method of photographing prominences in full sunlight." *Princeton University Archives.*

> 3 Wheatland Ave.
> Dorchester, Feb. 26, '88.
>
> Prof. E. C. Pickering
>
> Dear Sir:
>
> If convenient to you I shall be pleased to be at the Observatory every Saturday, commencing March 3, and continuing until June, from one to eleven P.M. Such an arrangement would be very satisfactory to myself, as I greatly wish to learn the work of the Observatory.
>
> Truly yours
> Geo. E. Hale

Edward C. Pickering.
Niels Bohr Library.

College Days at M.I.T.

Hale's first published paper, "The New Astronomy."
The Beacon, July 1889.

Vol. I.   JULY, 1889.   No. 7.

THE BEACON

A JOURNAL DEVOTED TO

PHOTOGRAPHY

IN ALL ITS PHASES.

CONTENTS:

	PAGE
THE COMING CONVENTION	145
NOTES	147
RETOUCHING.	
LIGHTING IN COPYING.	
BLUE LANTERN SLIDES.	
AWARD TO DR. R. L. MADOX.	
W. H. WALMSLEY.	
VARNISHING LANTERN SLIDES.	
IMITATION NOT ART.	
THE BLAIR CAMERA CO.	
DEUTSCHE GELATINE FABRIKEN.	
PHOTOGRAPHIC SOCIETY OF JAPAN.	
SYSTEMATIC DEVELOPMENT	150
VISITING THE STUDIOS	151
DALTONISM, OR COLOR BLINDNESS, BY W. H. DAVIES	152
WORDS FROM THE WATCH-TOWER, BY WATCHMAN	154

	PAGE
DARKROOM JOTTINGS, BY R. P. HARLEY	156
COLLODION EMULSION, BY R. MILLIKEN	157
FIGURE IN LANDSCAPE PHOTOGRAPHY, BY GRAHAM BALFOUR	159
MEETINGS OF SOCIETIES	162
CHICAGO LANTERN SLIDE CLUB.	
PHOTOGRAPHIC SOCIETY OF CHICAGO.	
PHOTOGRAPHIC SOCIETY OF PHILADELPHIA.	
ITALIAN PHOTOGRAPHIC SOCIETY.	
CHICAGO CAMERA CLUB.	
THE NEW ASTRONOMY	164
THE EDITORIAL TABLE	165
ANSWERS TO CORRESPONDENTS	167
UNITED STATES PHOTOGRAPHIC PATENTS	168
DARKROOMS OPEN TO TOURISTS	168

PUBLISHED MONTHLY AT $1.00 PER YEAR.   SINGLE COPIES, 10 CENTS.

THE BEACON PUBLISHING COMPANY,
TRIBUNE BUILDING,
CHICAGO, ILL., U. S. A.

Entered at the Postoffice at Chicago as second-class mail matter.

HENRY O. SHEPARD & CO., PRINTERS, CHICAGO.

THE NEW ASTRONOMY.

Two hundred and fifty years ago, when the discoveries of Galileo had shattered the popular belief in an immaculate solar surface and opened a new field of investigation outside of the earth, such an immense distance separated our planet from the sun that any knowledge of the chemical constitution of our light center seemed forever unattainable. Today the diameter of the earth's orbit is as great as ever, but a new instrument, even more wonderful and more powerful than the "optick tube" of the Florentine, has found, in the vibrations of the luminiferous ether, a method of analysis to which distance is no obstacle. Attached to the telescope, it brings to view gigantic gas streams upon the solar surface, never seen before its invention, except at the rare occurrence of a total eclipse ; with it their velocity is measured and their composition determined ; it tells us that iron, magnesium, calcium, hydrogen and a score of other elements exist in a state of vapor in the solar atmosphere ; it measures the velocity of rotation of the sun about its axis, and reveals the nature of the spots and faculæ. Applied to the study of the stars, it discovers their composition and velocity of motion toward or away from the earth ; it explains the sudden appearance of new stars in the heavens ; it tells us that the nebulæ are vast masses of glowing gas, and helps to unravel the mystery of comets. In fact, the New Astronomy, which has for its object the study of the physical and chemical nature of the heavenly bodies, owes its existence to the discovery and development of that most recent instrument of astronomical research, the spectroscope.

Since the memorable experiments of Kirchoff and Bunsen in 1859, when the prism first disclosed to their astonished gaze the secrets of the sun, a new science has sprung up, which offers problems not to be solved by the astronomer alone, but only by the combined skill of the astronomer, the physicist and the chemist. The votary of the new astronomy does not necessarily concern himself with the calculations of Newton or Laplace, but, watching in his laboratory the vibrations of the molecule in the Bunson flame or electric arc, he registers visible and even invisible phenomena upon the photographic plate, and then, comparing his work with the photographic record of stellar or solar vibrations, he is able to reason with clearness and certainty upon the constitution of the heavenly bodies. The fascinating questions of the new science have attracted astronomers and physicists in all parts of the world to the study of celestial phenomena, and in the short interval of thirty years the most astonishing progress has been made. A glance at some of the results obtained in that time may serve to illustrate the manifold interests of the subjects involved.

As has already been said, it was in 1859 that Kirchoff and Bunsen first laid the foundation of solar chemistry by the discovery of glowing sodium vapor in the atmosphere of the sun. The first step taken the advance was easy, and discovery followed discovery in rapid succession. Many of the elements were found to have bright lines in their spectra corresponding to dark lines in the spectrum of the sun, and the work of comparison was rapidly pushed forward by the Heidelberg chemists. With the first certainty of relation between bright and dark lines Kirchoff instantly saw the need of a new theory of the solar constitution, and for the cool, habitable globe imagined by Herschel in pre-spectroscopic days as the nucleus of the sun be substituted an incandescent sphere, hardly conducive to long life among the creatures of the elder astronomer's fancy. Surrounding this white hot mass he placed the elements, transformed by the intense heat into vapor, which filter out from the continuous spectrum of the central mass the rays corresponding to their particular rates of vibration, and leave to terrestrial observers a long band of color written over with hieroglyphs easily translatable to the careful student. This theory was at once accepted by all, and time has not overthrown the reasoning of the far-seeing German scientist.

Later the same kind of work was carried on by Augström at Upsala, and the coincidences more certainly established by an apparatus of higher dispersion, using a "grating"—a glass plate ruled with some thousands of parallel straight lines to the inch, instead of a train of four prisms such as was used by the earlier observers. Augström's map of the solar spectrum is a model of painstaking skill, and it is still to be found in every book on spectrum analysis.

The powers of the spectroscope were now firmly established, and it is not surprising that in 1866 Dr. Huggins turned his great reflector to the stars, and, by carefully keeping the star image stationary on the slit of a spectroscope placed at the focus of the instrument, he succeeded in photographing the spectra of the stars, and by a series of visual comparisons with the bright lines in the spectra of metals volatilized by the electric spark, he demonstrated the presence of many of the elements in the fiery atmospheres of these inconceivably distant suns. GEORGE ELLERY HALE.
(To be continued.)

THE EDITORIAL TABLE.

"An honest critic is the author's friend."

L. M. Prince & Brother, Cincinnati, Ohio, send us a neatly gotten up and profusely illustrated catalogue and price-list of 200 pages. It seems to include everything that a photographer can possibly require, and the arrangement and method of numbering the items have evidently been devised with a view to enable their customers to make out their orders with the least trouble.

The book finishes up with some thirty-six pages of "How to Make Pictures," including negatives, printing, transparencies, enlargements, photo-micrographs, etc. The instructions generally are sound and satisfactory ; but surely the time has gone by, or ought to have gone, for recommending the addition of alum to the fixing solution ; and equally antiquated is the supposition that it is necessary to mess with "translucine." Thanks to the Eastman Company for their stripping and transparent films, which have forever put an end to that *bête noire* of the photographer. There are also several pages of standard and selected formulæ that will be found useful.

To J. M. Forbes our best thanks are due for a beautiful lantern-slide of a "Newhaven Fishwife." It is simply perfect in every detail, although printed from a negative made from a cabinet print which we recognize as one of a set made by Ross & Thompson, of Edinburgh, certainly twenty-five years ago.

The average illustrations of most of the photographic journals are not of much value as object lessons, and many of them are positively injurious, as they are regarded by many who know no better, as examples of what should be the highest aim of the photographer. When, therefore, we come across anything that we consider in the highest degree meritorious, such as the picture by J. F. Ryder, in *Wilson's Photographic Magazine* of June, we like to direct attention to it. The picture is very aptly entitled, "The Future," and possesses that dreamy, thoughtful look that fits the title admirably. Composition, light and shade, and expression are so perfect that the highly tech-

The 8-inch telescope and attached spectroscope that Sir William Huggins used to photograph the spectra of stars in 1866.
Sir William and Lady Huggins, An Atlas of Representative Stellar Spectra (London: William Wesley and Son, 1899).

Jules Janssen, whom Hale first met at Meudon Observatory in 1886. In 1868 Janssen first observed solar prominences outside an eclipse (almost simultaneously with Norman Lockyer).
Niels Bohr Library.

Robert Ball's drawing of a prominence from visual observation. The disc represents the relative size of the earth.
Robert Ball, Story of the Sun (London: Cassell, 1893).

Sunspot of March 5, 1873, drawn by Samuel P. Langley, whose magnificent drawings of spots were an inspiring influence on Hale.
Langley, The New Astronomy (Boston: Houghton, Mifflin, 1889).

College Days at M.I.T.

Kenwood and Yerkes

Before going to M.I.T. Hale had become engaged to Evelina Conklin of Brooklyn, New York, who, like the Hale family, spent her summers in Madison, Connecticut. They were married the day after he graduated. On their honeymoon they visited the Lick Observatory on Mount Hamilton above San Jose, California. Years later Hale recalled that visit:

I shall never forget my first sight of the great Lick telescope, designed by my old friend Ambrose Swasey. It was in 1890, after two or three years work in solar research and opportunities to see most of the largest telescopes in this country had partially prepared me for this memorable experience. As our stagecoach wound its way up Mount Hamilton and our view spread wider and wider over the Santa Clara Valley my enthusiasm for the site of the Lick Observatory rose accordingly. But the climax was reserved for the evening, when we entered the great dome.

Darkness had fallen, and no light except that of the little sputtering spark on Keeler's spectroscope revealed the interior. Far above our heads, outlined against the sky through the opened shutter, the long tube reached up toward the heavens. At its extremity the 36-inch object glass, gathering the light from the remote nebula at which the telescope was directed, concentrated it in a sharp image 57 feet below. Here, after passing through the slit of the spectroscope, it was analyzed into its component lines. Their position, when measured with reference to the standard lines of the comparison spark, gave with unequalled precision the composition and the radial velocities of the planetary nebulae. Meanwhile the driving-clock carried the great tube steadily from east to west, compensating for the earth's axial rotation and holding the small nebula accurately in place on the narrow slit.

The impression made by this striking demonstration of what modern engineering, in the hands of a master, can do for astronomy has been intensified by time. It had been strengthened at the Lick Observatory by discussions with such leading astronomers as Keeler, Campbell, Burnham and Barnard, with whom I was to be intimately associated in later years, and by the realization that the work on nebulae we saw in progress was and still remains one of the classics of research. Within a few months I had ordered from Warner and Swasey the mounting of the 12-inch telescope of the Kenwood Observatory.[4]

At the Lick Observatory Edward Holden, the director, had offered Hale the chance to attach his spectroheliograph to the 36-inch telescope, then the largest in the world. But, disappointed by the solar "seeing," Hale decided to return to Chicago to develop his plan of photographing the prominences with a new 12-inch refracting telescope made by John Brashear. On June 15, 1891, the telescope was dedicated in the Kenwood Observatory behind the Hale house at 4545 Drexel Boulevard. In 1892, using the H and K lines of calcium in the ultraviolet, Hale photographed the bright calcium clouds or "flocculi" and the prominences all around the sun's limb for the first time, and thus proved the success of his new instrument.

In 1892 he was appointed Associate Professor of Astrophysics at the new University of Chicago. That summer he learned of the availability of two 40-inch lenses at the firm of Alvan Clark in Cambridgeport, Massachusetts, and saw the chance of obtaining a larger telescope for the study of the sun. He wanted one "that would carry powerful spectroscopes and spectroheliographs, and give a larger image of the sun suitable for the study of the structure of spots, flocculi, and prominences."[5] But how, he asked himself, could it be obtained? To whom could an appeal be made — in the name of the new University — for a really adequate observatory? In "The Beginnings of the Yerkes Observatory" he tells the story of the summer of 1892 and how his dream was realized:

I had not successfully answered this question in the summer of 1892, and as I cast my fly for trout from an Adirondack canoe I fear that my mind wandered from the attractions of the sport to the still greater charms of solar research. At any rate, when I left the mountains for a week to attend the meeting of the American Association [for the Advancement of Science] at Rochester, the much desired observatory was dominant in my thoughts. There I met [Edwin] Frost, who was just losing a German accent, Eliakim Moore, coming to Chicago from Yale, and other close friends of later years. And there I found the clue to my problem.

One hot evening we were sitting in front of the hotel, trying to keep cool. I was in a receptive mood, and my ears readily caught a tale that Alvan G. Clark was telling to a group about him. It seems that in 1889, when the Lick Observatory was beginning to create its world-wide reputation, the people of Southern California, then in the uproar of a land boom, felt that their fair territory might advantageously profit by the example of James Lick. So a worthy citizen offered to a local educational institution land then valued at fifty thousand dollars. This was ample warrant, in the judgment of the hour, for ordering a pair of 40-inch glass disks, which in the course of three years had been successfully made by Mantois in Paris. But unfortunately the land-bubble had meanwhile burst, the gift was worthless, and Mantois was vainly seeking payment of the sixteen thousand dollars at which the disks were valued. Here was a great opportunity, said Clark, for someone to get a larger telescope without loss of time. He had tested the glass and found it perfect, and nothing would please him more than to figure a 40-inch objective.

I may add parenthetically that the 40-inch telescope was to have been erected on Mount Wilson. Our subsequent occupation of this site, after Professor Hussey's telescopic tests had shown it to be more favorable than any other place he visited, was a pure coincidence.

It goes without saying that Clark's story gave me food for thought. I returned to the Adirondacks, packed my fishing tackle, and hastened to Chicago. After consulting my father, whose interest in my project was very keen, I visited several men who might conceivably be willing to provide for the telescope. But no one had the money to spare. A few days later I made another fruitless round of visits in the city. At noon, somewhat discouraged, I called at the Corn Exchange Bank to see Mr. Charles Hutchinson, then, as now, the enthusiastic friend and supporter of every such effort. After explaining my object, I asked for suggestions. "Why don't you try Mr. Yerkes?" he replied. "He has talked of the possibility of making some gift to the University, and might be attracted by this scheme." So I went at once to President Harper, then at the threshold of his tremendous task of building the University of Chicago. After a few questions he heartily approved of the attempt, asked me to write out a statement of the plan, and sent it to Mr. Yerkes. A reply came back asking us to call on him. We did so, and before the interview was over Mr. Yerkes asked us to telegraph for Clark, with whom he made a contract for the 40-inch objective. I remember with pleasure Dr. Harper's enthusiasm as we left the office. "I'd like to go on top of a hill and yell!" he cried.

Hale's arguments had appealed to Yerkes's vanity, as he saw the chance to glorify his name forever and gain a different reputation from that he had made for himself in Chicago. However, the financial problems were not so easily solved. Hale continued:

The University, initiated chiefly as an undergraduate school, was beginning [in 1892] to feel the influence of Harper's large and scholarly ambitions. The doors of its single building were about to open to those who boldly negotiated the generous mud of the "campus," but its problems were many and various....

It was natural that the embryo Observatory, consisting of a 40-inch objective in the process of figuring, should share in these difficulties. Indeed as I look back on President Harper's strenuous task, I realize that nothing short of his over-whelming optimism could have induced him to favor an undertaking which could play but little part in his educational scheme. With limitless plans for a great university, but without the lecture halls, dormitories, libraries, laboratories, and museums needed for the thousands of students that peopled his dreams; without the large faculty required for instruction and research; and without the many millions essential to meet current bills, it is no wonder, however, that he could not see except in Mr. Yerkes, any source of funds for a large and expensive observatory, situated far from the University campus and devoted almost solely to research. Mr. Yerkes, on the contrary, not appreciating the general problem, thought that Mr. Rockefeller's millions should immediately build, equip, and maintain the Observatory he had initiated. After some persuasion, however, he ordered the 40-inch mounting from Warner and Swasey and a stellar spectrograph from Brashear, but there he stopped and refused absolutely to do more.

Many vigorous efforts failed to break the deadlock, and I decided to go abroad for some months, partly for study at the University of Berlin and partly for an attempt to photograph the corona without an eclipse from some high mountain. Before leaving I requested the then University architect, Henry Ives Cobb, who was close to Mr. Yerkes, to let me know if he saw the least hopeful sign. While in Berlin, in December 1893, I received a letter from Cobb saying he thought that an interesting design might make some appeal. We had entertained not even a distant hope of securing a complete observatory building from Mr. Yerkes: a tower and dome for the 40-inch were all he could even talk of. But I determined to make a bold stroke for an adequate building, and accordingly prepared a comprehensive plan. Of this I heard nothing until one day the following Spring, when one fine day in Florence I received from Cobb a large roll of blueprints, showing my plan worked out as it now appears in brick and mortar. My delight was so great that I could hardly leave the blueprints behind me even when visiting picture galleries!

But our troubles were not over, as Mr. Yerkes was not yet ready to proceed with the work. After attempting—of course without success—to photograph the corona with a special spectroheliograph from the summit of Mount Etna in July 1894 I returned to Chicago, only to find that Mr. Yerkes was unwilling to build until the University would guarantee to maintain the Observatory on an adequate scale. We had been overwhelmed by enterprising real estate dealers with offers of tracts of land in various parts of Illinois and Wisconsin, and even the site finally selected had to be freed from encumbrance. I called for aid from three good friends of the Observatory, who promptly responded. But the perplexities of the University, hard pressed for funds in this period of serious financial depression, were not so easily overcome. Finally I guaranteed that the operating expenses would not exceed a very small sum during the first year, which meant that I must raise in addition money to pay salaries and to meet other expenses. At last the work of construction was begun in 1895.[6]

On October 21, 1897, the Yerkes Observatory was dedicated at Williams Bay, Wisconsin. It was based on a revolutionary principle, and, as Hale noted, it was "in reality a large physical laboratory as well as an astronomical establishment" where "all kinds of spectroscopic, bolometric, photographic and other optical work would be done in its laboratories."[7]

Here he gathered a small but devoted staff that included future astronomical leaders, and he encouraged visits from foreign astronomers, eager to use the superior facilities. He continued too the observation of sunspot spectra begun at Kenwood and designed the Rumford spectroheliograph to be attached to the 40-inch. With this powerful tool he found dark hydrogen flocculi on the sun and investigated the calcium flocculi at different levels to gain knowledge of the complex circulatory processes in the sun. From the sun he turned also to other stars, as he undertook (with Ferdinand Ellerman and John Parkhurst) a study of the spectra of the late-type, low-temperature red stars, known as Secchi's fourth type. These stars showed certain marked similarities to sunspot spectra and had never before been photographed. The observations were difficult, and fifty years later the noted astronomer Otto Struve pointed out that the results could be duplicated with only a very few of the world's largest telescopes, despite the tremendous developments in photography.

Meanwhile Hale had been pushing the astrophysical cause in other ways. On his triumphal tour in 1891 he had been welcomed by leaders in astronomy and physics in England and on the Continent. On his return, with their endorsement, he founded a journal called *Astronomy and Astrophysics* with William Payne, editor of the *Sidereal Messenger*. In 1895, with James Keeler as joint-editor, he formed the separate *Astrophysical Journal,* with an international board of editors that included astronomers and physicists from England, France, Germany, Italy, and the United States. This is still the leading journal in its field.

At the Yerkes dedication in 1897, a congress of astronomers was held. In 1898 a meeting preliminary to the founding of an astronomical society was held at Harvard, and in 1899 the first meeting of the new society was held at Yerkes. At first Hale, fearful that astrophysics in its embryonic state might be overlooked, insisted that it should be called the American Astronomical and Astrophysical Society. Simon Newcomb became the first president, with Hale and Charles Young as vice-presidents. In 1914, when astrophysics was more generally accepted, the name was changed to the American Astronomical Society.

Before the question of the name was decided, however, Hale and Newcomb had an amusing exchange of letters that reflected the difference in the viewpoints of the old and the new astronomy. Thus Hale wrote to Newcomb:

If, as you remark, we were all working in so ancient and honorable a science as astronomy, everyone would undoubtedly be willing to adopt the simple and effective name of "American Astronomical Society." But this does not appear to be the case.... The experiment of bringing together both the astronomical and physical phases of astrophysical work which has been carried on for the last four years in the *Astrophysical Journal,* may fairly be said, I think, to have met with some measure of success. If I am not mistaken, it has done something to stimulate the interest of physicists in astronomical problems, and it ought to have had some effect in convincing astronomers of the importance of applying in their own investigations the principles and methods which guide the physicist in the laboratory. I am unable to see why we should not do all we can to bring into harmonious and cordial cooperation all investigators who are concerned with the problems which interest us so deeply. We are certainly far from wishing to any way circumscribe the field of those who are working with the well-founded methods of the older astronomy. On the contrary, we hold that the dignity and importance of these researches can be in no wise impaired by public recognition of the apparent fact that astrophysics, because of the physical nature of many of its problems and the special methods devised for attacking them, has fairly acquired the right to occupy a place of its own, allied on the one side with astronomy and on the other with physics.... I cannot avoid the belief that the best interests of science in this country will be met by developing together these three branches (i.e., astronomy, astrophysics, and allied departments of physics) rather than by keeping them apart.[8]

The 36-inch refracting telescope at Lick Observatory.
Library of Congress.

23 Kenwood and Yerkes

This newspaper account of the dedication ceremonies at Kenwood Physical Observatory goes on to quote one of the speakers, John A. Brashear, the telescope maker:

A few years ago somebody commenced a correspondence with me by the name of George E. Hale. From the method of his writing, from his correspondence, I judged him to be a man of about 45, and to be certainly well up in astronomical business, and, strange to say, I had not heard of him through any of the books. One day, after I had furnished this same gentleman, living in Chicago, quite an amount of apparatus, there stepped into my shop a young fellow. I looked at him. He seemed to be very quiet and modest, and I thought: Well, there is another fellow wants a job as an apprentice. [Applause.] He came up to me and introduced himself as Mr. Hale, of Chicago. Well, if somebody had taken a baseball bat and hit me very, very hard I should not have been any more surprised. I took a liking to the fellow at once, and if he had been a lady I should have liked to have proposed, but he wasn't, and I was married anyway.

Chicago newspaper clipping, June 16, 1891.
Niels Bohr Library.

IN BEHALF OF SCIENCE.

Interesting Exercises Last Night at the Dedication of Kenwood Observatory.

An Admirably Equipped Institution Presided Over by George E. Hale.

Celebrated Astronomers Attend the Dedication and Commend the Plans.

CHICAGO IN SCIENCE.

Among the young men of Chicago there is one, Mr. George E. Hale, who would rather be a great astronomer than be President. At least, that is the way he expresses his love for the branch of science to which he has resolved to devote his life. Last evening the dream of years was realized for him, when the dedication of Kenwood Observatory took place.

KENWOOD OBSERVATORY.

Young Mr. Hale, who is the son of W. E. Hale, President of the Hale Elevator Company, graduated recently from Boston Polytechnic Institute, and at once turned his attention to a broader study of physical astronomy than he had been able to pursue at college. His study had already brought him in contact with some of the foremost men of the day in the line of physical astronomy, and his subsequent travels and correspondence extended his acquaintance.

Chicago newspaper clipping, about 1895.
Niels Bohr Library.

HE STUDIES THE SUN.

Original and Interesting Researches of Prof. George E. Hale.

SOLVING A DIFFICULT PROBLEM.

By His Work the First Photograph of Solar Prominences Was Made— His Latest Scientific Device.

To be an able astronomer at twenty-four is something; to have acquired a special knowledge of a special subject in science that is rare even among scientific men is something more; but to be a discoverer, to be a man with origination, with singleness and fixity of purpose, to be facile princeps among experts in a widely covered field at that very youthful age, is to be one in thousands, a genius in a century. Such is Professor George E. Hale, of this city, the director of the Chicago University observatory. Professor Hale's remarkable and interesting discoveries in solar observation alone entitle him to a prominent place in the world of science, a world more aristocratic than any nobility on earth. But when his scientific success is coupled with

PROFESSOR GEORGE E. HALE.

his youth they surpass much that is recorded in that marvelous book, the history of modern astronomy. If the future of his work may be gathered from its past Professor Hale will certainly add more to the stock of human knowledge about the sun than any other one man.

Photographing the Sun.

It is difficult—nay, impossible—to explain clearly the methods of his work to one not thoroughly familiar with solar spectroscopy and the uses of the spectroscope. But the results can be stated generally. He discov-

George Hale in his laboratory at the Kenwood Observatory.
Niels Bohr Library.

Photograph of a prominence, July 9, 1894, taken with the spectroheliograph attached to the Kenwood 12-inch telescope.
Photo by Ferdinand Ellerman, Hale's assistant. Hale Observatories.

Potsdam Observatory near Berlin, where Hale spent "many profitable hours" during the winter of 1893.
David P. Todd, Stars and Telescopes (Boston: Little, Brown, 1899).

George Hale's wife Evelina, his brother William, his father, and his sister Martha photographed by Hale in the Adirondacks in 1892.
Niels Bohr Library.

The mounting for the 40-inch telescope built for Yerkes Observatory on display at the Columbian Exposition in Chicago, 1893.
Niels Bohr Library.

Charles T. Yerkes.
News photo. Niels Bohr Library.

Alvan G. Clark and Carl A. R. Lundin with the 40-inch lens that had taken them five years to prepare for the Yerkes telescope.
Yerkes Observatory, University of Chicago.

29   Kenwood and Yerkes

THE BIGGEST IN THE WORLD.
A Contrast in Modern Civilization.

YERKES BREAKS INTO SOCIETY

Street-Car Boss Uses a Telescope as a Key to the Temple Door

AND IT FITS PERFECTLY

AT THE OTHER END.

THE INHABITANTS OF THE MOON ALSO ENJOY THEIR FIRST GLIMPSE OF THE YERKES TELESCOPE.

FEAST OF WISE MEN.

Yerkes Gives a Banquet to Visitors and Citizens.

HARPER IS TOASTMASTER

Scientists at Kinsley's Laud the University Telescope.

DONOR SAYS IT IS FOR ALL.

Ryerson Laboratory Is Inspected by Wondering Guests.

ORATION BY PROF. NEWCOMB

Chicago newspaper clippings, 1892–1897, on the establishment of the Yerkes Observatory.
Niels Bohr Library.

MENU

Bluepoints

Consomme Dumas

Hors d'oevres

Fillet of Black Bass au Vin Blanc
Cucumbers

Tenderloin of Beef Moderne

Cardinal Punch

Breast of Partridge   Lettuce Salad

Glaces   Cake

Coffee
Sherry   Sauternes   Champagne
Cognac   Cigars

TOASTS

Toast Master, Hon. Egbert Jamieson

The University and Astronomy — Mr. Ferd. W. Peck

The University and the College — Mr. D. K. Pearsons

The Work of an Observatory — Mr. E. C. Pickering, Harvard University

Sister Observatories — Mr. J. K. Rees, Columbia University

The Naval Observatory — Mr. William Harkness

The Neighbors of the Yerkes Observatory — Mr. George C. Comstock, University of Wisconsin

Instrument Making — Mr. J. A. Brashear

The Ryerson Physical Laboratory — Mr. A. A. Michelson, the University of Chicago

The Astronomical Staff — Mr. George E. Hale, the University of Chicago

Menu for one of the banquets during the dedication festivities, October 18-22, 1897.
Niels Bohr Library.

John D. Rockefeller, founder of the University of Chicago, and William R. Harper, its first president, were among the dignitaries at the Yerkes dedication.
University of Chicago Archives.

Group at the dedication.
Yerkes Observatory, University of Chicago.

1. E. E. Barnard
2. C. H. Rockwell
3. George F. Hull
4. Colton (?)
5. Kurt Laves
6. Frank W. Very
7. E. B. Frost
8. Henry M. Paul
9. Ernest Fox Nichols
10. F. R. Moulton (?)
11. Ephraim Miller
12. Father John Hedrick
13. John M. VanVleck
14. Milton Updegraff
15. William R. Brooks
16. F. L. O. Wadsworth
17. H. C. Lord
18. F. H. Seares
19. George W. Hough
20. W. H. Collins
21. Caroline E. Furness
22. Mrs. Pickering
23. Mrs. Hale
24. Miss Cunningham
25. Alva A. Lyon
26. Carl A. R. Lundin
27. J. A. Parkhurst
28. John A. Brashear
29. G. W. Ritchey
30. C. H. McLeod
31. Father J. G. Hagen
32. Charles Lane Poor
33. J. K. Rees
34. Miss Mary W. Whitney
35. F. P. Leavenworth
36. Henry S. Pritchett
37. James E. Keeler
38. A. G. Stillhamer
39. Hugh L. Callendar
40. George W. Myers
41. Charles L. Doolittle
42. E. C. Pickering
43. A. W. Quimby
44. Asaph Hall
45. Albert S. Flint
46. M. B. Snyder
47. W. W. Payne
48. Carl Runge
49. Winslow Upton
50. George Kathan
51. G. D. Swezey
52. George E. Hale
53. N. E. Bennett
54. George C. Mors
55. F. Ellerman
56. W. J. Humphreys
57. Henry Crew

Identifications made about January 1, 1923, by F. Ellerman, E. E. Barnard, S. B. Barrett, J. A. Parkhurst.

George Hale in his office at Yerkes Observatory.
Yerkes Observatory, University of Chicago.

E. F. Nichols, H. M. Goodwin, E. E. Barnard, E. B. Frost, G. E. Hale, F. Ellerman, F. Schlesinger, J. A. Parkhurst, G. W. Ritchey, and an unidentified colleague on the steps of Yerkes Observatory, August 14, 1898. Friends and astronomers were frequent visitors in summertime.
Yerkes Observatory, University of Chicago.

PROF. HALE AND ASSISTANT ADJUSTING THE SPECTROSCOPE.

Chicago newspaper, 1897.
Niels Bohr Library.

The Rumford spectroheliograph designed by Hale was attached to the 40-inch telescope in 1899. With it Hale and his colleagues were soon obtaining spectacular results. Chicago newspaper clipping, 1899.
Niels Bohr Library.

Calcium flocculi observed by Hale for the first time as luminous clouds above the sun's disc in photographs taken with the Rumford spectroheliograph. Shown also are solar prominences erupting at the sun's limb. *Yerkes Observatory, University of Chicago.*

In the winter Yerkes Observatory was often isolated by heavy snow.
Niels Bohr Library.

Hale strolling with his children, Margaret and Bill, along the wooded shore of Lake Geneva near the Observatory.
Niels Bohr Library.

Simon Newcomb, first president of the American Astronomical and Astrophysical Society.
National Academy of Sciences. Biographical Memoirs, XVII. Washington, D.C.: 1916.

A draft title page for the new astrophysical journal showing its international board of editors.
The Hale Papers.

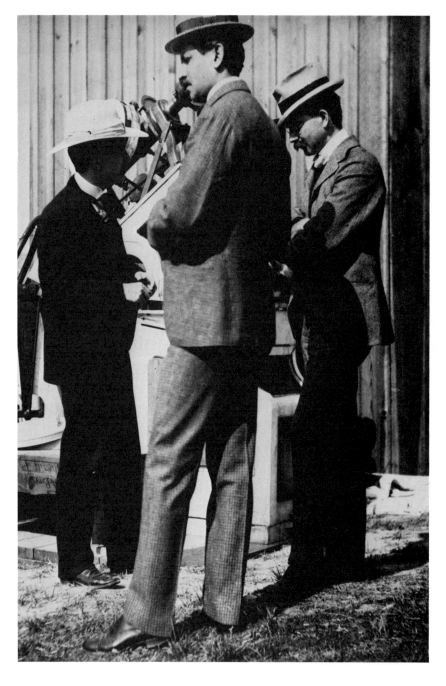

The three college friends, Hale, Noyes, and Goodwin, were among the scientists who gathered at Wadesboro, North Carolina, for the solar eclipse of May 28, 1900.
The Smithsonian Institution.

Samuel P. Langley, secretary of the Smithsonian Institution, preparing to observe the eclipse at Wadesboro using a 5-inch equatorial telescope. *The Smithsonian Institution.*

Mount Wilson

In 1896 Hale had persuaded his father to provide the disc for a 60-inch reflecting telescope with which stellar spectra could be photographed "on so large a scale as to permit the study of their chemical composition, the temperature and pressure in their atmospheres, and their motions with that high degree of precision" that could then be reached only in the case of the sun.[9] William Hale had offered this disc to the University of Chicago on condition that funds be found to mount it. George Hale tried repeatedly without success to fulfill this condition. The disc remained in the Yerkes basement.

The opportunity finally came in 1902 when the Carnegie Institution of Washington was founded by Andrew Carnegie to "encourage investigation, research and discovery in the broadest and most liberal manner, and the application of knowledge to the improvement of mankind."[10] Two years later, after overcoming countless difficulties and gambling $30,000 on the outcome, Hale founded the Mount Wilson Solar Observatory on a peak above Pasadena, California, and there in 1908 the 60-inch reflector finally went into action.

Many years later, in 1929, Hale described the founding in a letter to Mrs. Walcott, the widow of Charles D. Walcott, secretary of the Executive Committee of the Carnegie Institution, who had been one of his most ardent supporters:

I vividly recall my intense interest in reading the first announcement of the establishment of the Carnegie Institution of Washington. It was briefly told in a Washington dispatch to a Chicago newspaper, but the provision of a large endowment solely for scientific research seemed almost too good to be true. For ten years I had been devoting much of my energy, at the heavy cost of my investigations, to the difficult and often unpleasant task of raising funds.... The University could not have been expected to give more help than it did, as anyone closely acquainted with its origin and history will agree. I was thus bound to undertake the heavy task of raising funds or to forego the possibilities I seemed to see ahead.

These were nothing less than an effective union of astronomy and physics, directed primarily toward the solution of the problem of stellar evolution, but with equal consideration for the advantages to be gained by fundamental physics from such a joint study....

Without describing in detail the conditions that restricted the application of this plan at the Yerkes Observatory, I may add that my hopes for this branch of the University of Chicago had been only partially realized when I read the account of the endowment of the Carnegie Institution in 1902. The 40-inch refractor supplied only a fraction of the light needed for various astrophysical researches, and its type of mounting precluded the formation of the fixed solar and stellar images required for investigations possible only with powerful fixed spectrographs, radiometers and spectroheliographs, rigidly mounted under laboratory conditions. I had neglected no opportunity to round out our equipment, as the provision of the "heliostat house" in the Observatory building, the construction of the Snow telescope and its ill-fated predecessor, and the attempt to build a 60-inch reflecting telescope, with mounting designed to give a fixed stellar image (Ranyard's plan), sufficiently indicate. My father had purchased a 60-inch mirror disk, and provided so far as he could for its grinding and polishing by

[George] Ritchey, whose technical skill had been shown by his optical and photographic work. But my father was unable to do more than to initiate this plan, in the hope that another donor could be found to provide the funds necessary for the completion of the 60-inch telescope and its building.... After repeated failures, our hopes were at a low ebb in 1902.

Hale then describes the long and involved train of events that led to the founding of the Mount Wilson Observatory under the auspices of the Carnegie Institution. He tells too of his first visit to Wilson's Peak in June 1903, his enthusiasm over the possibilities of research with new and powerful instruments in such a climate, and of his decision to make an expeditionary trip from Yerkes with a small coelostat for testing the conditions more thoroughly.

On Feburary 29, 1904, accompanied by one carpenter and two burros loaded with tools and supplies, I started for the summit of Mount Wilson, where Messrs. Staats and Holmes of Pasadena, who owned the summit as members of the Pasadena and Mount Wilson Toll Road Company, had kindly agreed to let us occupy the "Casino" as a dwelling and to set up instruments near it.

The "Casino" was an ancient log cabin, formerly used by summer visitors at Strain's Camp, but long since fallen into decay. All of the glass in its many windows had been smashed and a great hole in its roof offered an ample view of the heavens as we lay in our cots during the first nights. While the carpenter jacked up and repaired the roof I extracted the remaining fragments from the windows and fitted them with new glass. When nails, putty, or other supplies were needed I walked down the nine mile trail, rode my bicycle to Pasadena, and returned in the same way. One of the delights of the downward trip was to shoot the "slides," which were cut through the heavy covering of "chapparal" as short-cuts between zigzags of the trail. Needless to say my costume was that of a California pioneer and I took home with me abundant quantities of the powdered dust collected on the slides and trail.

At the "Casino" we did our own cooking and fared sumptuously on such delicacies as flapjacks and bacon. The season was one of little rain and luckily we were well inclosed before the first heavy storm struck us. I had decided to provide one luxury, a stone fireplace and chimney, which proved our salvation when real winter descended with its furious blasts of snow....[11]

Hale's reaction to the primitive conditions on the mountain where all supplies had to be transported by burro or mule was a surprise to some of his colleagues who saw in his background no preparation for such a life. His close friend and successor, Walter Adams, writes of his insight, courage, and enthusiasm and his unexpected reaction to the novel conditions.

Apparently combined with a deep-seated love of nature in every form was the spirit of the pioneer, whose greatest joy is the adventure of starting with little and taking an active personal part in every phase of creation and growth. To both of these inborn characteristics of Hale, Mount Wilson in 1904 offered a rich field and scope for their full employment.[12]

Mount Wilson Solar Observatory was founded on December 20, 1904. Soon afterwards, in 1905, the first photograph of a sunspot spectrum ever made was taken with the Snow telescope, an instrument that was essentially a solar telescope, fed by a coelostat, devised to accommodate larger, more powerful spectrographs than could be attached to the 40-inch refractor.

By this time a small laboratory had been built on the mountain. Here spectroscopic results, obtained with the Snow telescope and other instruments, could be analyzed and compared with laboratory results obtained under controlled conditions. In this way, the significant observation was made by Hale, Walter Adams, and Henry Gale that those spectral lines that are strengthened in sunspots are exactly the lines that are strongest in low-temperature sources, such as the electric arc and furnace. Thus it became evident that sunspots are cooler than other regions of the solar disc, as Hale had long suspected.

In 1908, in the hope of overcoming the temperature problems that had plagued the low-lying Snow telescope, he had designed and built a 60-foot tower telescope with a 30-foot spectrograph in an underground pit. With photographic plates sensitive to red light he was able to detect vortices in the hydrogen flocculi in the vicinity of sunspots. This led him to the hypothesis that the widening of lines in spot spectra might be due to the presence of intense magnetic fields in spots. He became convinced that the splitting was due to the effect, first observed by Pieter Zeeman in 1895, when a line spectrum is placed in a strong magnetic field. In 1908 he proved it as he compared his observations with the doubling of lines observed with a powerful electromagnet in his Pasadena laboratory. This observation of magnetic fields in sunspots was Hale's greatest discovery. It was also the first discovery of an extraterrestrial field. The significance of Hale's work was noted by Robert Woodward, president of the Carnegie Institution of Washington, who wrote, "This is surely the greatest advance that has been made since Galileo's discovery of those blemishes on the sun."[13]

Today it is believed that the vortical motion, observed by Hale above sunspots, has nothing to do with the generation of the magnetic field in these spots. It is held by the majority of astronomers that a clearly marked vortex appears in only a rather small percentage of spots. Nevertheless it was Hale's hypothesis on the cause of vortices above spots that led him to explore the possibility of finding a magnetic field. Therefore, the methods he used and the way he reached his result are invaluable as a part of the historical record. (See "Solar Vortices and Magnetic Fields," page 166.) This discovery led, in turn, to his recognition of the reversal of spot polarities with the spot cycle and to the formulation of a fundamental polarity law. In this law Hale proved the 23-year interval between successive appearances in high latitudes of spots of the same magnetic polarity.

Meanwhile he was considering the puzzling question: Is the sun itself a magnet? The 150-foot tower telescope with a 75-foot vertical spectrograph, designed to obtain the spectral resolution needed to measure the sun's general field, was completed in 1912. Preliminary observations with this instrument indicated that the sun has a dipole field with a strength of about 20 gauss. Further observations in the 1930s indicated a field of approximately 4 gauss. But the results were inconclusive. It was not until 1952 that H. D. and H. W. Babcock, using a specially designed instrument in the Hale Solar Laboratory in Pasadena, developed the first reliable method of measuring magnetic fields on the sun's surface.

In 1908, twelve years after his father had given him the disc, the 60-inch reflecting telescope, then the largest in the world, was set up on Mount Wilson. At last, with its great light-gathering power, the first steps could be taken in the photographing of stellar spectra on a scale that might eventually approach the great dispersion available for the study of the solar spectrum. The way was prepared for an understanding of stellar evolution that would be realized only when knowledge of atomic processes gained in earthly laboratories could be applied to the interpretation of the nature of stars and nebulae and when, in turn, knowledge derived from studies of those "enormous crucibles," the stars, could be applied on earth. In *The Study of Stellar Evolution*, published in 1908, Hale described the immense possibilities.[14] By 1915 the 60-inch was justifying his hopes for it, but the results showed only how much more might be gained with a still larger telescope.

Hale's father often said that even as a boy George always wanted things "yesterday." This continued to be true all through his life. In 1906, with the success of the 60-inch still uncertain, he had described the possibilities of a 100-inch telescope to John D. Hooker, a Los Angeles man in the hardware business. It would, he said, give two and a half times as much light as the 60-inch, seven times as much as any other telescope then in use for stellar astronomy. It would "enormously surpass all existing instruments in the photography of stars and nebulae, giving new information on their chemical composition and the temperature and pressure in their atmospheres."[15] With his talent for convincing wealthy men of the urgent need for supporting his dreams, he persuaded Hooker to donate the funds for a 100-inch disc. The telescope was built with Carnegie funds, largely as a result of Andrew Carnegie's visit to Mount Wilson in 1910. It was completed at the end of 1917, and was soon hard at work solving problems that up to that time had seemed insoluble. In 1920, Albert Michelson, using it with his interferometer, measured the velocity of light; and Francis Pease and John Anderson performed the extremely difficult feat of measuring the diameter of the giant red star Betelgeuse and found it to be an astounding 300 million miles.

Perhaps one of the most perplexing problems in astronomy is that of the scale of the universe. When the 100-inch was set to work, it was hoped that a definite answer might be found. At the end of 1923 Edwin Hubble identified a Cepheid variable, one of those "yardsticks of the universe," in a spiral nebula, and thus found the key to its distance. This result, as Allan Sandage points out, "proved beyond question that nebulae were external galaxies of dimensions comparable to our own. It opened the last frontier of astronomy, and gave, for the first time, the correct conceptual value of the universe. Galaxies are the units of matter that define the granular structure of the universe."[16] Without the 100-inch telescope this breakthrough in our knowledge of the universe would have been impossible at that time.

Hale has been called the "master builder" in recognition of his role in the building of large telescopes. But he was also a builder of institutions. He was, as Hunter Dupree has noted, a major statesman of his era—"one of the first prototypes of the high-pressure, heavy-hardware, big-spending, team-organized scientific entrepreneurs. Would that all who followed him on this path had his technical competence, his clarity of scientific objective, and his breadth of view."[17]

George Hale on his first visit to Wilson's Peak in 1903 to check observing conditions.
The Hale Observatories.

The view of Mount San Antonio from Wilson's Peak.
*Photograph by Ferdinand Ellerman.
The Hale Observatories.*

Mount Wilson

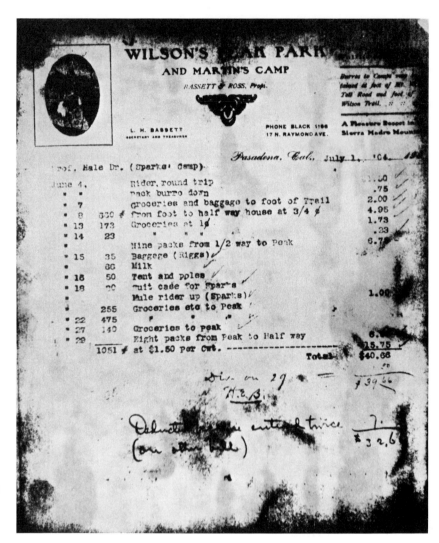

The cost of carting supplies by burro is reflected in this bill for the month of June 1904.
The Hale Papers.

The "Casino" which served as a shelter in the early days on the mountain.
The Hale Observatories.

It took a full day for the burro train to make the slow and rough trip up the narrow trail to the top of Wilson's Peak.
The Hale Observatories.

49   Mount Wilson

GROUP AT MONASTERY (1905)

Miller　　　　　Abbot　　Hale　　　Ingersoll　　Adams　　Barnard　　　　Backus
(construction　　　　　　　　　　　　　　Ellerman　　　　　　　　　　　(assistant)
superintendent)

The Hale Observatories.

50   George Ellery Hale 1868-1938

Hale on Mount Wilson with Ferdinand Ellerman, whose ability as a photographer and skill with instruments had been indispensable to Hale since the days at Kenwood Observatory.
The Hale Observatories.

The building of the great stone piers for the Snow telescope, the first telescope to be installed on the mountain. It was ready for use by the summer of 1905.
The Hale Observatories.

Los Angeles Express, December 21, 1904.

PLUM FOR MT. WILSON

Big Grant for Solar Observatory by Carnegie Institution

Will Expend $300,000 in Its Equipment and Provide for Its Maintenance—Dr. Hale in Charge

Today a telegram was received from Washington stating that the Carnegie Institution for the promotion of science had made a grant of $150,000, for use in 1905, for the purpose of establishing a solar observatory on Mt. Wilson. As the observatory is to be an integral part of the Carnegie Institution, its further equipment and maintenance will be provided for by subsequent grants. It is estimated that the equipment of the observatory will cost not less than $300,000.

Dr. George E. Hale, director of the Yerkes observatory of the University of Chicago, has been appointed director of the new solar observatory. In the new work he will be assisted by G. W. Ritchey, superintendent of instrument construction and assistant professor of practical astronomy at the Yerkes observatory, and by Ferdinand Ellerman and Walter S. Adams, instructors in astrophysics at the Yerkes observatory. Professor Ritchey will have charge of the shop in Pasadena for the construction of instruments needed by the new observatory, and Messrs. Ellerman and Adams will take part in the observational work on Mt. Wilson. Other members of the staff will be appointed later.

Plans for the establishment of a solar observatory have been under consideration by the trustees of the Carnegie Institution since 1901. In that year Dr. S. P. Langley recommended that an observatory be built on a high mountain for the purpose of determining whether the sun's heat is perfectly constant in amount, or undergoes changes corresponding with the great changes observed in the sun spots and other phenomena of the sun's surface. The recommendation was referred to a committee consisting of Professor E. C. Pickering of Harvard, Prof. Lewis Boss of Albany, Prof. George E. Hale of the Yerkes observatory, Prof. S. P. Langley of Washington, and Prof. Simon Newcomb of Washington. The committee reported favorably on the project, and suggested that if the Carnegie Institution were sufficiently interested in the matter it would be advantageous to appoint a committee for the special purpose of preparing a complete plan for a solar observatory. The duties of this committee also were

Sir William Huggins. The portrait painted by John Collier is at the Royal Society.
The Scientific Papers of Sir William Huggins (London: William Wesley, 1909).

Sir William Huggins wrote to Hale on January 30, 1907,

It must indeed be of untold satifaction [*sic*] to you to be at the head of so magnificent an Institution for combined astronomical and laboratory work! It seems to me almost marvellous that within a single life-time, my first taking a simple spectroscope into an Observatory has borne such magnificent fruit in both hemispheres.
The Hale Papers.

90, UPPER TULSE HILL, S.W.

Jan. 30th, 1907.

Dear Dr. Hale,

I should have replied before to your very kind letter of the 4th, but I have been suffering from a sudden and rather sharp attack of Influenza. I am now practically well again.

It is true that I have a pressing invitation from the Secretary of the Carnegie School to be present at the opening, I gave him no hope that we should be able to go, but he wished me to defer a final answer. Great as would be the pleasure of seeing my American friends in their own country, I am sure that it would not be prudent for me to undertake a journey involving so much fatigue.

It is indeed some consolation that we may indulge the hope of seeing you on this side of the water. To your visit we shall look forward eagerly and with pleasure.

It must indeed be of untold satifaction to you to be at the head of so magnificent an Institution for combined astronomical and laboratory work! It seems to me almost marvellous that within a single life-time, my first taking a simple spectroscope into an Observatory has borne such magnificent fruit in both hemispheres.

I am delighted to hear that you are to add to your wonderful equipment a reflecting telescope larger than any yet attempted. I am sure that you will do great things with it.

Lady Huggins joins with me in sympathy with you on account of Mrs Hale's long ill-health. We unite in your hope for a better state of health. We shall be glad to have the photograph of your children, who we are glad to hear are well.

With united kind remembrances to Mrs Hale and to yourself,

I remain, yours sincerely,

William Huggins

The presence of intense magnetic fields in sunspots was suggested to Hale by vortices in the hydrogen flocculi as shown in this photograph taken September 9, 1908.
The Hale Observatories.

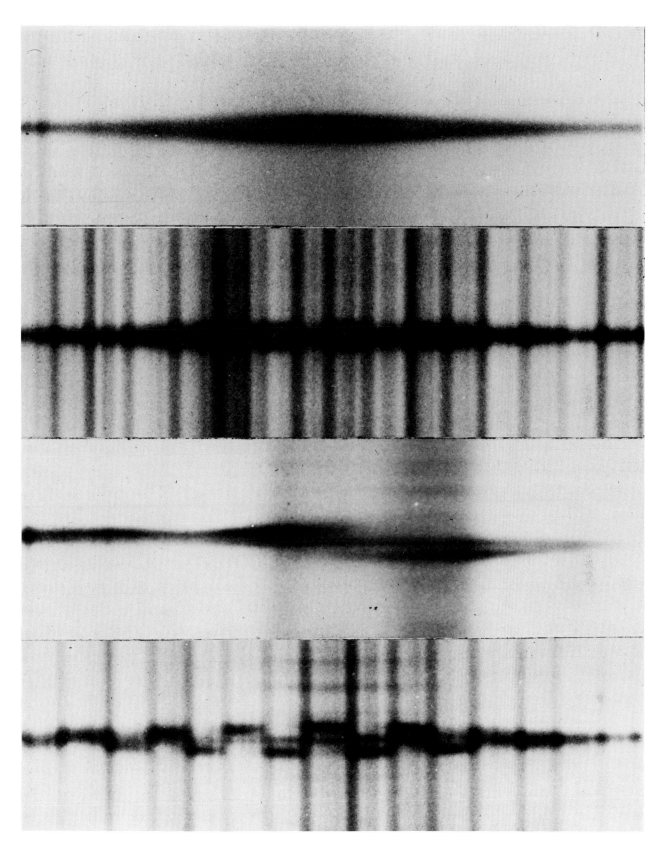

Spectra showing the Zeeman effect, the splitting of lines in a magnetic field.
The Hale Observatories.

Pieter Zeeman.
Burndy Library.

Pieter Zeeman in a letter to Hale, November 25, 1908:

After balancing the evidence for and against the probable existence of magnetic fields in sun-spots... I very decidedly came to a favourable opinion....

We live in a beautiful period of physics and astrophysics. You may look back upon the past year with immense satisfaction. I hope you may demonstrate one day the magnetic field in a spiral nebula.

The Hale Papers.

I would suggest that they are in reality the quartets, polarized as given in the figure and with rather large separation of the two middle components.

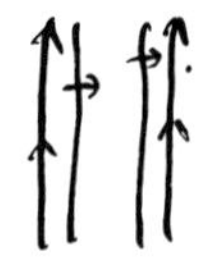

Your Addendum received just now, I read with extreme interest. The bit of evidence given by the asymmetrical triplet 6302.71 is most beautiful.
I thank you very much for your extremely kind accompanying letter. The duplicates of the lost photographs I received a few hours ago and a few days ago the splendid photographs of the sun with the vortices rotating in opposite direction. For all these marks of kindness I thank you very sincerely.
We live in a beautiful period of physics and astrophysics. You may look back upon the past year with immense satisfaction.
I hope you may demonstrate one day the magnetic field

3.

of a spiral nebula.
With kind regards, yours very sincerely
P. Zeeman.

Pasadena newspaper clipping, undated.
Niels Bohr Library.
Andrew Carnegie and Hale in front of the 60-inch dome on Mount Wilson.
The Hale Observatories.

CARNEGIE ON MT. WILSON

FIRST TRIP.

MR. CARNEGIE UP IN WORLD.

Climbs Mt. Wilson to Repay Debt to Science.

Towering Pine Tree There Catches Fancy.

Rain Shuts Out Stars He Hoped to See.

Andrew Carnegie slept last night in a little cabin 6000 feet above the sea, after taking the finest mountain trip he has ever enjoyed in his life, as he expressed himself upon his arrival on Mount Wilson yesterday. An immense yellow pine supports one side of the cabin. The monarch of the mountain immediately caught his fancy.

"Nine hundred years old," he said, passing his hand over the rough bark. "What a pity that a man could not have a life of such length. It would increase his opportunities so much."

Mr. Carnegie's pine tree is 120 feet in height and seventeen feet in circumference. In spite of the cold and damp, he stayed out on the porch long to admire it.

Standing there, he talked of his hopes for the scientific work on the mountain—the work that he has supplied the money for and is now there to inspect.

WHAT MAY BE DISCOVERED.

"We do not know what we may discover here," he said. "Franklin had little idea what would be the result of flying his kite. But we do know that this will mean the increase of our knowledge in regard to this great system of which we are a part.

"Mr. Hale has discovered here 1600 worlds about one of the stars which were not known before. We have found helium in the sun, and after finding it there we discover it in the earth. It all goes to show that all things are of a common nature. What may be the relation of our discoveries

60   George Ellery Hale 1868-1938

Anstruther Davidson, George Hale, James H. McBride, John Muir, Henry Fairfield Osborn, John D. Hooker, James Scherer, and Andrew Carnegie at the Hotel Maryland in Pasadena during Carnegie's visit.
The Hale Observatories.

Mrs. Carnegie, her daughter Margaret, Evelina Hale, and her daughter Margaret with Carnegie on a tour of the laboratory at Mount Wilson.
The Hale Observatories.

Carnegie's note to Hale of November 27, 1911, reflects his concern over Hale's health. Hale had suffered a nervous breakdown the previous year. *The Hale Papers.*

New York, November 27, 1911

My dear Frend,-

Delited to read your long note this morning; not too long—every word tells, but pray show your good sense by keeping in check your passion for work, so that you may be spared to put the capstone upon your career, which should be one of the most remarkable ever livd.

Ever yours,

Andrew Carnegie

George E. Hale, Esq.,
  Palace Hotel,
    San Francisco.

Regards &
Madam & Mine

Photograph of the lunar crater Copernicus taken by Francis Pease with the 100-inch reflector in 1919, for many years the finest lunar photograph in existence.
The Hale Observatories.

Spiral nebula in Canes Venatici, Messier 51, photographed with the 60-inch telescope.
The Hale Observatories.

The same spiral photographed with the 100-inch telescope shows the greater detail observable with the larger instrument.
The Hale Observatories.

Albert Einstein.
Burndy Library.

Zürich. 14.X.13.

Hochgeehrter Herr Kollege!

Eine einfache theoretische Überlegung macht die Annahme plausibel, dass Lichtstrahlen in einem Gravitationsfelde eine Deviation erfahren.

Am Sonnenrande müsste diese Ablenkung 0,84" betragen und wie $\frac{1}{R}$ abnehmen (R = Entfernung vom Sonnen-Mittelpunkt).

Es wäre deshalb von grösstem Interesse, bis zu wie grosser Sonnennähe helle Fixsterne bei Anwendung der stärksten Vergrösserungen bei Tage (ohne Sonnenfinsternis) gesehen werden können.

Albert Einstein to Hale, October 14, 1913.
*My very dear Colleague:

A simple theoretical consideration makes plausible the assumption that light-rays experience a deviation within a gravitational field

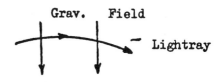

On the edge of the sun the deflection should be 0.84" and as $\frac{1}{R}$ diminish (R = Distance from the center of the sun)

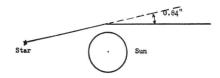

Therefore it would be of the greatest interest to know within how great a proximity to the sun bright stars could be seen by daylight [without a solar eclipse] with the help of the largest magnification.

Upon Prof. Maurer's recommendation I write to you, and beg you to let me know what the chances are to obtain results with the present-day equipment and your vast experience in this field.

*Addressed to Prof. George E. Hale, Pasadena, Calif., by Dr. Albert Einstein, Zürich, Switzerland.
The Hale Papers. Translation by Erwin Morkisch.

November 8th, 1913.

Professor Dr. A. Einstein,

　　　Technische Hochschule,

　　　　　Zürich, Switzerland.

Dear Professor Einstein:-

　　　　　I have dealyed replying to your kind letter of October 14th until I could consult Director Campbell of the Lick Observatory, who I knew to be interested in the problem you describe. He writes me that he has undertaken to secure eclipse photographs of stars near the sun for Doctor Freundlich of the Berlin Observatory, who will measure them in the hope of detecting differential deflections. Doubtless he will send you further particulars, as I requested him to communicate directly with you.

　　　　　I fear there is no possibility of detecting the effect in full sunlight, for the following reasons:

　　　　　1. The sky increases greatly in brightness near the sun, even under good observing conditions. I cannot now say at what distance bright stars would be visible, but will have observations made to determine this.

　　　　　2. On Mount Wilson the best definition of the solar image is obtained only for about an hour in the early morning. Hence the atmospheric refraction, changing rapidly with the hour angle, would be

Professor Dr. A. Einstein. -2-   11-8-1913.

a troublesome obstacle.

3. It would be necessary to measure the differential change in distance of the star from the sun's limb, which would be difficult because of the low precision of micrometer settings on the limb, and the large distance (probably much beyond the range of an ordinary micrometer).

The eclipse method, on the contrary, appears to be very promising, as it eliminates all of these difficulties, and the use of photography would allow a large number of stars to be measured. I therefore strongly recommend that plan.

In a short time, as soon as some additional data are available, I wish to ask your opinion regarding the theory of the general solar magnetic field which I have recently detected by observation of the Zeeman effect.

Believe me, with kind regards to Professor Maurer,

Yours very sincerely,

Hale replied to Einstein on November 8, 1913, "I fear there is no possibility of detecting the effect in full sunlight....
"The eclipse method, on the contrary, appears to be very promising... and the use of photography would allow a large number of stars to be measured."
The Hale Papers.

Charles G. Abbot relaxing with his cello on Mount Wilson. For many years Abbot spent the summer on the mountain studying the solar constant
Photograph by Margaret Harwood taken in 1915.

Hale and Alicia Mosgrove who shared an enthusiastic friendship from their first meeting at the home of John D. Hooker. Here in the Hooker garden they enact a scene from a script they had written.
Niels Bohr Library.

Sir James Jeans (center) with Mount Wilson astronomers Walter Adams and Edwin Hubble during his visit to the Observatory in the spring of 1931. *Photo courtesy of Mrs. Walter S. Adams.*

National and International Affairs

All his life Hale was interested first and foremost in research. But to achieve his goals in astronomy, in science, in the humanities, he realized early in life that he must divert some of his energies to the less appealing tasks of organization. In 1902 he was elected to the National Academy of Sciences. From the beginning he felt that the leading scientific academy in the United States should accomplish much more than it was doing if it was "ever to occupy its proper position in the scientific world" and "acquire a commanding influence of a favorable character, favorable alike to the development of research and the public appreciation of science."[18] To change its hoary ways and widen its influence, he proposed an increase in the membership, with an emphasis on younger, more forward-looking scientists. To broaden its outlook, he urged that the membership be expanded to include branches such as engineering and archaeology. To establish it on a sound foundation, he proposed a central headquarters in Washington. To enhance its international position, he urged programs of cooperation, especially in astronomy. In 1893 he had helped to arrange an international astronomical congress in connection with the Columbian Exposition in Chicago. In 1904 an International Exposition was to be held in St. Louis, and he proposed that a committee be formed to organize the "International Union for Cooperation in Solar Research" under Academy auspices. A number of European astronomers came to the first meeting held in St. Louis, including Henri Poincaré, who was made vice-president, while Hale was chosen president. The Union was formally organized at Oxford in 1905. At a large meeting on Mount Wilson in 1910 its aims were expanded to include all branches of astronomy.

Still Hale was dissatisfied. In 1912, as the Academy was to celebrate its fiftieth anniversary the following year, he summed up his ideas in a 9-page letter to Charles D. Walcott, its vice-president. He noted,

> Some of the members of the National Academy are entirely content with the Academy as it exists today, and others insist that its chief function is to confer honor upon those whom it elects to membership. With regard to the last point, it seems to me that the honor conferred by such election must depend in large degree upon the part played by the Academy in the advancement of science. If we can really make the Academy a strong factor in promoting scientific advance in this country, we can surely increase the honor of membership in it.[19]

To this end he would devote much of his energies in the years ahead. (See "The Future of the National Academy of Sciences," page 177.)

In 1914, as war broke out in Europe, the outlook for international exchange dimmed. But with the war Hale saw the chance to increase the Academy's usefulness in other ways. Out of that vision came the National Research Council, born during the war and, with the war's end, the new building and central headquarters of which he had so long dreamed. In a rough draft written some years later, "War Services of George E. Hale," he described the result:

In 1915, soon after the sinking of the "Lusitania," I proposed that the National Academy of Sciences, of which I was Foreign Secretary, should offer its services to the President of the United States. In April 1916, after the submarine attack on the "Sussex," I renewed the suggestion, which was then adopted by the Academy. President Wilson at once accepted the offer and asked that the Academy should bring into cooperation governmental, educational, industrial and other research agencies, primarily in the interest of national defense, but with full recognition of the duties that must be performed in the furtherance of scientific and industrial progress. I was appointed chairman of the committee that planned the National Research Council for the purpose in the summer of 1916. As soon as the preliminary scheme had been completed, I sailed for Europe with Dr. William H. Welch, president of the Academy, to learn how the men of science of France and Great Britain had served in the war.

On my return to the United States the Research Council was organized and I was elected Chairman. I retained this position, with headquarters in Washington, until May 1919, when I resigned, and was made permanent Honorary Chairman.

The work of the Research Council during the war covered a wide field, involving the study of problems in physics, chemistry, engineering, medicine, biology, geology, geography, agriculture etc., as well as the initiation of the sound-ranging and meteorological services of the Army, the psychological testing of troops, and the development of devices for the detection of submarines. In 1918 I received from President Wilson an executive order, authorizing the war duties of the Research Council and providing for its perpetuation in time of peace, under the Federal charter of the National Academy of Sciences.

I also assisted in obtaining a gift of five million dollars from the Carnegie Corporation for a building and endowment for the National Academy and the Research Council; subscriptions to purchase a site near the Lincoln Memorial in Washington, where the building has since been erected; and a grant from the Rockefeller Foundation of one hundred thousand dollars a year for the establishment of National Research Fellowships in physics and chemistry (since increased to three hundred thousand dollars a year to provide also for fellowships in other subjects).

As Foreign Secretary of the Academy, I prepared in 1918 a plan for the organization of an International Research Council, which was approved by the Council of the Academy and adopted in slightly modified form at conferences held under the auspices of the Royal Society in London and the Paris Academy of Sciences in Paris in the autumn of that year. The International Astronomical Union, initiated at this time, was designed to unite into a single body the various international astronomical organizations which formerly dealt with special branches of the subject.[20]

Charles D. Walcott, who supported Hale's efforts to found the Mount Wilson Observatory.
The Smithsonian Institution.

150-Foot Steel Tower Will Be Erected on Mt. Wilson's Summit

Rushing Work to House New Solar Instruments; Savants Are Coming in August

That science may gaze through powerful lenses at the marvels of the solar system, Mt. Wilson's summit has been reared 150 feet higher. And another marvel was accomplished in the rearing. For scientists attached to the Mt. Wilson Solar Observatory of the Carnegie Institution have erected a vibration-proof tower of steel on which will be mounted delicate mechanisms swaying to the will of man the most powerful reflecting mirrors.

Construction work on the tower has been rushed during the past three months in order that the work should be completed by August. Next month there will come to Southern California and to Mt. Wilson the savants of the world, men who have achieved fame in the study of the sun, the moon and the stars, all of them members of the International Union for Co-operation in Solar Research.

The tower, an idea of Prof. George Ellery Hale, director of the observatory, is so constructed as to enable scientists to make observations during periods of the day not now covered by other instruments in operation. It will also expose more minutely the details of phenomena under study in the sun. It will magnify spectra to a degree admitting more definite deductions and conclusions by studying scientists.

A slender tower of pressed steel mounts to a height of 150 feet above the solid cement foundations. The inner tower is protected from the wind by the outer and a dome of steel erected on the outer tower performs the same office for the delicately adjusted instruments mounted on the inner. Thus protected these instruments are rendered perfectly stable and will project the sun's rays unwaveringly into the pit beneath the tower floor.

The combined spectograph and spectoheliograph of seventy-five foot focal length are to be mounted in the pit. The pit is ten feet in diameter and seventy-eight feet deep. It is walled with concrete and has a winding staircase running from the tower floor to the bottom.

The coelostat for the new tower is now being finished in the Santa Barbara street workshops of the observatory at Pasadena. The mirrors are also being ground and will soon be ready for mounting.

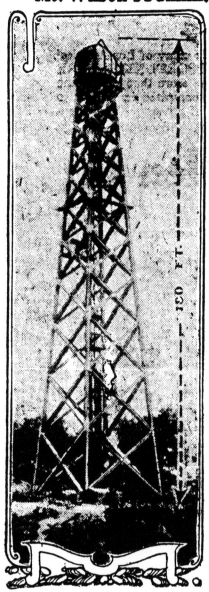

The 150-foot tower telescope, completed in time for the Solar Union meeting on Mount Wilson in 1910. *Pasadena Star-News. July 1910.*

The Hale Observatories.

FOURTH CONFERENCE
INTERNATIONAL UNION FOR COOPERATION IN SOLAR RESEARCH

August 30 - September 3, 1910

Mount Wilson

1. ELLERMAN
2. H. C. WILSON
3. ST. JOHN
4. LARKIN
5. TOWNLEY
6. V. M. SLIPHER
7. FOWLE
8. COBLENTZ
9. FROST
10. IDRAC
11. PUISEUX
12. HARTMANN
13. KÜSTNER
14. SLOCUM
15. HAMY
16. KNIGHT
17. WOLFER
18. FATH
19. RYDBERG
20. HEPPERGER
21. FOX
22. HAUSSMANN
23. CORTIE
24. TURNER
25. RUSSELL
26. KAYSER
27. ADAMS
28. MILLER
29. AMES
30. BACKLUND
31. KONEN
32. PICKERING
33. FOWLER
34. LAMPLAND
35. HALE
36. BELOPOLSKY
37. DESLANDRES
38. SCHUSTER
39. CAMPBELL
40. RICCO
41. MRS. KAPTEYN
42. BOSLER
43. K. SCHWARZSCHILD
44. MC ADIE
45. KAPTEYN
46. MRS. FLEMING
47. WATSON
48. SCHLESINGER
49. HUMPHREYS
50. MADRILL
51. J. F. SANFORD
52. CHRETIEN
53. DE LA BAUME PLUVINEL
54. FABRY
55. ABBOT
56. HILLS
57. LARMOR
58. COTTON
59. DYSON
60. BARNARD
61. KING
62. NEWALL
63. PRINGSHEIM
64. LEUSCHNER
65. J. S. PLASKETT
66. GALE
67. CHANT
68. EVERSHEIM
69. ROTCH
70. W. MITCHELL
71. STRATTON
72. H. D. BABCOCK
73. RITCHEY
74. BRACKETT

Edward C. Pickering, director of the Harvard College Observatory, and Hale's close friend, Hugh F. Newall, professor of astrophysics at Cambridge University, during the meeting on Mount Wilson.
The Hale Observatories.

Washington, April 26, 1916

My darling—

Our visit to the President was very satisfactory. Dr. Welch stated our general purpose and asked me to present the plan more in detail, which I did very briefly. The President saw the point, asked a few questions, and then requested us to proceed with the work at once. He preferred to have us act with only a verbal request, as the situation with Germany is so delicate that a letter from him might be misinterpreted. So this letter is confidential. We have a free hand to proceed as we choose, and I have been asked to become chairman of an executive committee in full charge of the work. Bill has also been made a member of this committee at my suggestion, and this will be a great help. Conklin and Noyes are also on it, and I can select any others I wish. But the responsibility will be a heavy one.

I have called a meeting of the committee in New York next Saturday and wired Bill to come on. My plans until June will have to be changed completely, I am afraid, as so much of the work must be done in the east. But while I cannot spend as much time as I expected in Chicago, I still expect to go to the University celebration and I want you to join me before then. As soon as I can map out the work with the Committee I will write you more definitely of my probable movements. Alas for the quiet days in the Boston libraries, to which I had looked forward so pleasantly! I fear they will be few and far between.

The situation has been changed by the fact that we are to go ahead with the work, war or no war. Of course not much can be done in the way of research unless a break comes, as the men would hardly drop their other work unless war seemed probable. But we must get all the universities and research institutions to offer their men and facilities in the event of a break, gather information regarding work to be done, register all those who are available for service, and do many other things in the interests of preparedness.

Margaret is still here, as she dreads as to go back I have kept her here as long as possible. Probably we will go back tomorrow, as there is much to be done in New York.

With much love from us both
Your loving
Husband

Hale to his wife Evelina, April 26, 1916, describing President Wilson's response to Hale's proposal on the formation of the National Research Council.
The Hale Papers.

Woodrow Wilson with Edward M. House, his friend and adviser. It was through "Colonel" House that Hale succeeded in obtaining Wilson's public approval of the National Research Council in July 1916.
Edward M. House Collection, Yale University Library.

Sir J. J. Thomson, the British physicist. Hale consulted with him in 1916 about scientific war research in England, and in 1918 regarding the International Research Council. Thomson is shown here with Frank B. Jewett in the laboratory of the Western Electric Company in 1923.
Western Electric Company. Courtesy National Academy of Sciences Archives.

William H. Welch, president of the National Academy of Sciences, who accompanied Hale to Europe in the summer of 1916.
Photo by Harris and Ewing. William H. Welch Medical Library, The Johns Hopkins University.

Hale's account of his war work.
The Hale Papers.

Hale's application for a special passport to attend the organizational meetings of the International Research Council in the autumn of 1918.
The Hale Papers.

DESCRIPTION OF APPLICANT.

Age: 50 years. Mouth: Gray Mustache
Stature: 5 feet, 8 inches, Eng. Chin: Square
Forehead: High Hair: Gray Thin
Eyes: Gray Complexion: Ruddy
Nose: Pointed Face: Long Oval
Distinguishing marks: Scar back 2nd finger left hand

IDENTIFICATION.

I, DeForest S. Mulvin, July 31, 1918, solemnly swear that I am a {native} citizen of the United States; that I reside at Pasadena Cal; that I have known the above-named George E Hale personally for 12 years and know {him} to be a native citizen of the United States; and that the facts stated in {his} affidavit are true to the best of my knowledge and belief.

DeForest S Mulvin
Bookkeeper (Occupation.)
Mt Wilson Observatory
(Address of witness.)
Pasadena Cal

Sworn to before me this 31 day of July, 1918 By B Zimmerman, Deputy
Chas. N. Williams, Clerk U. S. District Court, Southern District of California.

[SEAL.]

Clerk of the _____ Court at _____

Applicant desires passport to be sent to the following address:

George Ellery Hale,
1003 – 16th St
Washington
D.C.

National Research Council
Special Passport Requested

A duplicate of the photograph to be attached hereto must be sent to the Department with the application, to be affixed to the passport with an impression of the Department's seal.

The New York Times,
April 29, 1924.

The building for the National Academy of Sciences nearing completion early in 1924.
National Academy of Sciences Archives.

Figures of Galton, Gibbs, Helmholtz, Darwin, Lyell, and Faraday on one of the window panels of the National Academy building showing the founders of science from Greek to modern times.
National Academy of Sciences Archives.

COOLIDGE DEDICATES 'TEMPLE OF SCIENCE'

New Building of National Academy and Research Council Opened With Addresses.

PRESIDENT LOOKS AHEAD

Search for Truth in the Scientific Spirit Will Benefit Whole National Life, He Says.

Special to The New York Times.

WASHINGTON, April 28.—President Coolidge delivered the chief address today at the dedication of the building of the National Academy of Sciences and the National Research Council. The exercises attracted a large number of scientists.

President Coolidge, in his speech, recalled the beginning of scientific study in this country, said that we had made great progress, and that "this magnificent building now being dedicated to science predicts a new day in scientific research."

A. A. Michelson, President of the academy, presided, while the Right Rev. James E. Freeman, Bishop of Washington, delivered the invocation. John C. Merriam, Vice President of the academy, and Vernon Kellogg, Secretary of the council, also spoke briefly.

"If there be one thing in which America is pre-eminent," said President Coolidge, "it is a disposition to follow the truth. It is this sentiment which characterized the voyage of Columbus. It was the moving impulse of those who were the leaders in the early settlement of our country, and has been followed in the great decisions of the nation through all its history. Sometimes this has been represented by political action, sometimes by scientific achievements. On this occasion the emphasis is on the side of science.

"By science I mean the careful assembling of facts, their comparison and interpretation. Of those who are entitled to high rank in both our political and scientific life, perhaps Benjamin Franklin was the earliest and one of the most conspicuous examples. But it is the same spirit that has moved through all our life, which makes it particularly appropriate that our national Government should be active in its encouragement of the searching out of the truth in the physical world, and applying it to the wellbeing of the people, as it is interested in the searching out of the truth in the political world, with the same object in view.

C. Bascom Slemp, Albert A. Michelson, Charles D. Walcott, Vernon Kellogg, President Coolidge, John C. Merriam, Bishop James E. Freeman, and Gano Dunn at the Academy building on the day of its dedication, April 28, 1924.
National Academy of Sciences Archives.

Oil portrait of George Hale by Seymour Thomas. Three days after the dedication of the building, the Academy adopted a unanimous resolution that a portrait of Hale hang in the Academy:

as a permanent memorial and an adornment of the walls of the fine building which it owes in large measure to his unselfish and untiring efforts in furthering the material and intellectual interests of the Academy these many years, and whose preeminence in science, universally recognized throughout the scientific world, has added distinguished honor to the Academy.

National Academy of Sciences Archives.

Pasadena as a Cultural Center

The Mount Wilson Observatory was founded in 1904. In 1906 Hale became a trustee of Throop Polytechnic Institute, a Pasadena school with meager resources and diverse standards. It had "about 550 students, and taught a great variety of subjects, including elementary school, grammar school, high school, normal school, college, and some technological work, with considerable stress laid on manual training."[21] When Hale was asked by the trustees to discuss the possibilities of the school, he proposed that its character be changed entirely, so that it could become an institution of the first rank like M.I.T., only broader in outlook.

I emphasized the fact that it would be far better to do some one thing extremely well than to teach such a variety of courses in a mediocre way. They finally decided to convert the school into a high-grade institute of technology, and to begin with courses in mechanical and electrical engineering.[22]

In the same letter to James Scherer, its future president, Hale urged him to gamble on its future:

As you will see, the school does not amount to very much at the present time, but it seems to me that the right kind of President could easily make a great institution of it in the near future.... If it were a question merely of duplicating existing technical schools, I do not think the problem would be of exceptional interest from the standpoint of an educator. I believe, however, that it certainly seems feasible to educate men broadly and, at the same time, make them into good engineers. In my opinion, the man who works out some such plan will make a contribution of the highest importance to educational methods....

Hale also sent Scherer a copy of a recent paper published in the *Technology Review* that related to M.I.T. (see "A Plea for the Imaginative Element in Technical Education" on page 145). He pointed out that Throop, as a new school unfettered by tradition, would be able to carry out these wider goals more easily. It was a difficult, if challenging task. Yet, by 1920, Hale could write to Ernest Rutherford in England:

You will be glad to know that Throop College of Technology in Pasadena has just received a gift of one hundred and fifty thousand dollars for a physical laboratory from Dr. Norman Bridge. [Robert] Millikan who spends three months of each year at Throop College, has designed an excellent building, with complete arrangements for research which we hope to have developed there on a considerable scale in the course of time. Professor Arthur A. Noyes, who is an old friend of mine, has recently resigned the position he has held for the past thirty years at the Massachusetts Institute of Technology, where he was Director of the Research Laboratory of Physical Chemistry, and has established himself here permanently as Director of Chemical Research at Throop College, where a number of important investigations in chemistry are already under way. [Albert] Michelson has already arranged to spend much time here....[23]

In 1920 Throop College (its name was soon to be changed to the California Institute of Technology) was well on its way to becoming the important educational institution with a broad outlook of which Hale had dreamed. This dream was part of a larger plan for Pasadena as a cultural center.

In 1906 Hale learned from the railroad magnate, Henry Huntington, of his plan to give his magnificent collection of paintings and rare books to Los Angeles County. Hale, fearful of a scheme that had political overtones, began a campaign to persuade Huntington to consider instead the possibilities of a center in the humanities where scholars from the world over might come to do research, just as astronomers had come to Yerkes and Mount Wilson. Nearly twenty years later, after a lengthy, often discouraging correspondence and rare meetings, Hale, who had been made a Huntington trustee in 1919, presented him with a complete plan for the future development of a research library to be placed under the direction of a great and productive scholar, aided by a permanent corps of original investigators of the highest caliber. Shortly before his death in 1927 Huntington provided the endowment for the Henry E. Huntington Library and Art Gallery that contained, in essence, the kind of research center Hale had proposed many years before, and had so often described in glowing terms in the intervening years. One such letter to Huntington was written on May 11, 1914.

The powerful attractions of your pictures and library have fired my imagination, and set in motion a new train of ideas. I can't help feeling that with such rich and uniquely valuable material it would be a very easy matter to make your collection of real international importance, without greater expenditure than you may already contemplate. I have had some experience in starting international projects, and have learned that it is quite as easy to acquire the hearty cooperation and support of the best men in Europe as to carry out an enterprise of merely local significance. The one thing essential is to have a project so well worth doing that it appeals as strongly to Europeans as to Americans....

Your pictures include many of the finest specimens in existence of the work of the greatest artists, and would command wide attention wherever exhibited. Your library, the richest private collection in the world, contains material for much literary study. But in view of the existence of extensive collections of pictures and books abroad, and also in a few of our largest cities, some methods of increasing the attractions of your collections in the eyes of literary men and artists should be considered.

Here Hale launched into a romantic description of the home he envisioned for this magnificent collection of books and pictures. He proposed a building that would be a *perfect* copy, in Pentelic marble, of some Greek temple—a building that, like the great library of Alexandria, would become "the center of the scholarly interest and research of the civilized world."

Hale concluded:

Merely to make your collections accessible to the public would confer a great benefit. But to do this, and at the same time to establish a true international center, known throughout Europe and America as a unique attraction to every lover of art or of literature, would be so tremendously well worth doing that the increase of expenditure required to accomplish it would be repaid a hundred fold. With such a plan in prospect it would be easy to organize an advisory international board of the greatest living scholars and artists, whose advice and assistance could be obtained whenever required....[24]

To make Pasadena itself a place of distinction, Hale also worked on a city plan that included a scheme for highways and parks and for a group of public buildings, designed by leading architects. These included the City Hall, the Auditorium, the Public Library. For this work, he received the city's highest award, the Noble medal.

James A. B. Scherer, president of Throop from 1908 to 1920, at his desk in the Institute.
California Institute of Technology Archives.

Architect's drawing of the campus planned for Throop Polytechnic Institute in 1908.
The Hale Papers.

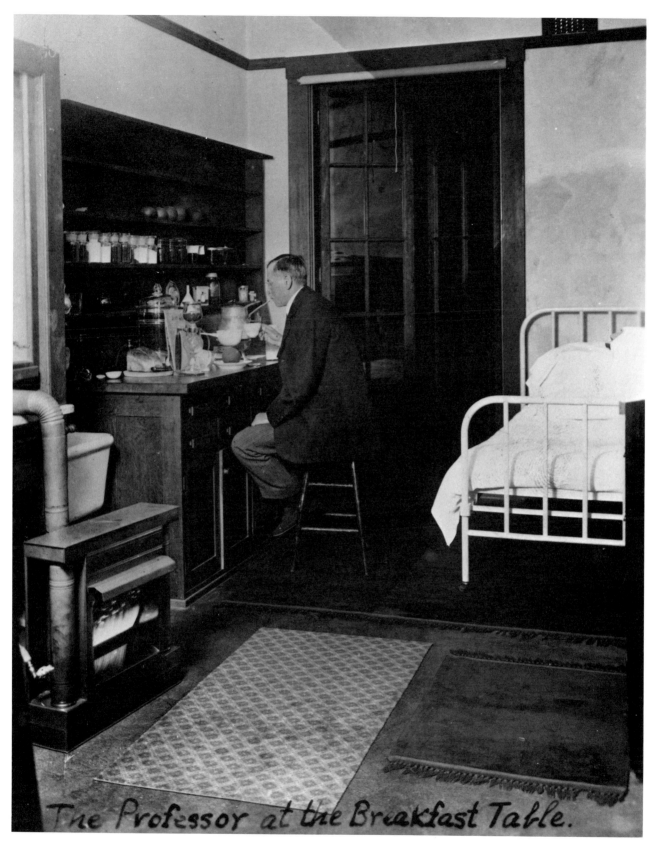

Arthur Noyes in his laboratory at Throop.
The Hale Papers.

Caricature of Millikan by Arthur Cahill in the Athenaeum at the California Institute of Technology. In 1921 Millikan became chairman of the Executive Council of the Institute.
California Institute of Technology Archives.

Hale, Arthur Noyes, and Robert Millikan in front of the Norman Bridge Physics Laboratory.
The Hale Papers.

View of the campus, 1960s.
California Institute of Technology Archives.

93   Pasadena as a Cultural Center

Mr. Henry E. Huntington
2 East 57th St.
New York City.

May 11, 1914

My dear Mr. Huntington:

Many thanks for your letter, which I found here on my return to Pasadena. I shall be glad if my suggestions prove to be of any service.

The powerful attractions of your pictures and library have fired my imagination, and set in motion a new train of ideas. I can't help feeling that with such rich and uniquely valuable material it would be a very easy matter to make your collection of real international importance, without greater expenditure than you may already contemplate. I have had some experience in starting international projects, and have learned that it is quite as easy to acquire the hearty cooperation and support of the best men in Europe as to carry out an enterprise of merely local significance. The one thing essential is to have a project so well worth doing that it appeals as strongly to Europeans as to Americans.

In 1894 I organized the Astrophysical Journal, an international review of spectroscopy and astronomical physics,

with reference to each other, and the best form of approach to give the finest architectural effect, etc. Choisy shows in his history of architecture how largely carefully the Greeks worked out all of these details, and how largely the success of the Parthenon and the other buildings of the Acropolis depended upon them. In addition to such practical points, on which a landscape gardener like Olmsted and an architect like Coolidge would be of the greatest value, there would be many other questions on which the advice of the leading Shakespearean scholars of England, the head of the Archaeological Service of the Egyptian Government in Cairo, the governing members of the German Oriental Society and the chief German Academies engaged in archaeological research, etc, would be no less important. If the time should come when you would care to have the assistance of such an international board, and I could help in any way by organizing it, I should be very glad to do so.

Sir Francis Younghusband, the distinguished British officer and explorer who conducted the expedition to Lhassa, has recently visited us at the Observatory. He was greatly impressed with Pasadena and its climate, which he recognized as much superior to that of the Riviera, where cold winds are so prevalent. In his opinion many Europeans will come here for the winter as soon as the Panama Canal is open, especially as the whole voyage can be made in southern waters. This would greatly facilitate the plan of making your collections of international importance.

I should have hesitated to offer these suggestions if they had involved a large increase of expenditure or a material change in your own plans. As you will see, they mean only slight modifications in detail, but I believe the result would be most satisfactory to you. Merely to make your collection accessible to the public would be a great thing. But to confer a great benefit, and to do this and at the same time to establish an international center of art and literature, known throughout Europe and America and a great attraction to every lover of art or of literature, would be so tremendously well worth doing that the moderate increase of expenditure required to accomplish it would be repaid a hundred fold. With such a plan in prospect it would be easy to organize an advisory international board of the greatest living scholars and artists, whose advice and assistance could be obtained whenever required.

Pardon me for writing at such length. The interest of the great possibilities got the better of me, and I now find it difficult to stop!

Believe me, with kindest regards to Mrs. Huntington,

Yours very sincerely,

Experience has taught me the advantage of gathering suggestions from such sources, as some entirely new point of view may result from them. All sorts of questions would require study, such as the best possible sites for buildings, their exact arrangement

(over)

2 EAST 57TH STREET

Dear Professor Hale —

Your letter of May 11th reached me as I was sailing for Europe and during the summer I have given the suggestions some thought. I am not ready to reply but it is quite possible that you have planted a seed. We returned last week from the lands of the "Far Flung Battle Line" and it is good to be back.

Mrs Huntington joins me in kind regards —

Sincerely yours,
H E Huntington

Oct. 5th 1914

Extracts from thirteen-page draft of letter from Hale to Henry E. Huntington, May 11, 1914, proposing a plan for the preservation and use of Huntington's books and paintings.
The Hale Papers.

Huntington replied to Hale on October 5, 1914, "... during the summer I have given the suggestions some thought. I am not ready to reply but it is quite possible that you have planted a seed."
The Hale Papers.

Huntington in the doorway of the Henry E. Huntington Library.
Henry E. Huntington Library and Art Gallery.

Aerial view of the Art Gallery and the Library. The Gallery was built by Huntington in 1910 as a country residence on the San Marino ranch.
Henry E. Huntington Library and Art Gallery.

The Hale Solar Laboratory and the 200-inch Telescope

By 1923, as he continued to be plagued by the ill health that had earlier helped to bring on two nervous breakdowns, Hale gave up the directorship of the Mount Wilson Observatory and built the Hale Solar Laboratory where he hoped to carry on his own solar research. Here he designed and built his first spectrohelioscope for the visual study of solar phenomena and the spectroheliograph with which he continued his investigation of the still puzzling problem of the sun's magnetic field. His enthusiasm over his observations of solar phenomena with the spectrohelioscope is reflected in a letter to his nephew, George Hale:

I am sitting before the fire in my Lab., thinking of the night when you and I discussed all sorts of questions before the crackling faggots at Winnetka. Five minutes ago I finished the enclosed article for the Proceedings of the National Academy, which your father may like to see.[25] It represents most of the work (which has kept me as busy as such an old codger can be) done since the day I got home. I never had a bullier time, not since the 3989 [Drexel Boulevard] days when we used to sally forth on Saturday mornings into the shop behind the house. But I must tell you what has been under way.

Imagine yourself on the "Nautilus," and suppose that all fish were just naturally invisible. But you have contrived certain peculiar windows, through one of which you can see all the sharks, through another nothing but whales, through another only the octopi or pusses. My "Nautilus" is hard aground, and I must look out through a sea of air instead of water. Beyond the air stands the sun, with wild and fantastic beasts roaming over its surface. They are of enormous size, sometimes reaching up to heights of four hundred thousand miles. And they are fearfully hot, made of hydrogen, helium, and calcium, in the form of gas so thin that it is only about a thousandth part as dense as the air you are breathing. Naturally, when you look at the brilliant surface of the sun with a telescope you do not see these beasts, because they are so thin that they don't cut out any appreciable fraction of the light. But if you had a window through which you could see nothing but things made of hot hydrogen, they would suddenly come into view. I have made such a window, and have been looking through it at these marvelous beasts much of the time since I came home. Their antics are astonishing, and though I knew something about them before (because I had a method of photographing them without seeing them), it is quite a different matter to see them diverting themselves like monsters of the deep. So you will understand why I have been kept very busy, and have had no time, like yourself, for defunct Egyptian bugs. It may not be so easy to bring them to life, as the enclosed letters from Dr. Lillie and Dr. Flexner indicate.

The new work has made it necessary for me to design several new devices and to work on some of them in my shop. This has also been great sport, and I have often wished that you and your father would be with me there,

Warm and merry, singing, laughing,
Shouting O, Kabibinoka,
You are but my fellow mortal!....[26]

At times, however, as he sat in his "Solar Lab.," Hale's thoughts flew back to his days at "Boston Tech." On one such day he wrote to his classmate and lifelong friend, Harry Goodwin, who by this time had become dean of the Graduate School at M.I.T.:

I had just finished reading your splendid article in the Technology

Review when the little book containing it and the programme with your welcome letter also arrived. To-day I have received the well-illustrated pamphlet on the Graduate School. What an immense amount of satisfaction you must get out of all these developments. And how far away Boylston St. and the old Rogers Lab. seem! Many vivid memories came into my mind as I read your article, of Charles and Sile and Cliffy and most of all of *you*. I don't get up as early in the morning as I did in the Dorchester days but I wish I had the chance to try it and to sally forth, leap on the ancient horse-car, and thrust my feet deep under the straw until time to hop off at Roxbury for the fine old walk to town.

But what would I have thought of great vacuum spectrographs and long focus concave gratings in those days? Perhaps you may remember me carrying around in my pocket the sketches of my first Kenwood Lab. and the blue prints of the 10 foot concave grating Brashear was making for me! I still rejoice in new and powerful instruments, but I can hardly duplicate now the wild thrill of excitement that came with my earliest and simplest equipment: that in the "Little Room" at 3989 Drexel of the early eighties. There is where I saw my first spectra, made oxygen, blew up hydrogen, mounted microscope specimens, made an induction coil, etc. And never since, I suppose, have I had *quite* such an exciting time. But perhaps distance lends enchantment. However, no magnification by time is necessary to enhance the real delight of our days together at the old Tech. And it is a fair question, I suppose, whether the students of to-day get any more satisfaction from the fine buildings and powerful equipment we have both slaved so long to bring into being. They do get much better and more advanced instruction, and my small stock of knowledge often makes me feel like thirty cents as compared with that of students here. Luckily far abler men than I are on hand to instruct them....[27]

If, however, Hale looked back during these years of his "retirement," he also looked forward. He had not forgotten the dreams that had led him to build three of the world's largest telescopes and led him now to dream of a still larger telescope that could penetrate more deeply distant and still unexplored realms of the universe. In 1928 forty years had passed since he had first urged the concept of an observatory as a physical laboratory on a skeptical astronomical world. By 1928 this concept was no longer questioned. But he was eager to apply it even further to the 200-inch telescope that he described in an article for *Harper's Magazine*, "The Possibilities of Large Telescopes." (See page 193.) He wrote: "Lick, Yerkes, Hooker and Carnegie have passed on, but the opportunity remains for some other donor to advance knowledge and to satisfy his own curiosity regarding the nature of the universe and the problems of its unexplored depths."[28] But who was this other donor to be?

When the article was published, he asked the editor to send a proof of it to Wickliffe Rose at the Rockefeller Foundation. At the same time he wrote a long letter to Rose in which he emphasized the progress attained by the 100-inch in the solution of "many fundamental, astronomical and physical problems beyond the reach of the 60-inch reflector." He pointed out the need for a 200-inch telescope with its greater light-gathering power in the attack on still more important problems, including the evolution of stars and nebulae and the constitution of matter "since the enormously greater range in mass, temperature, pressure, and density of the heavenly bodies present opportunities for discovery far beyond the possibilities of laboratory experiment."[29]

To Hale's surprise Rose responded immediately and expressed great enthusiasm for such a project. This led, in time, to the donation by the International Education Board of the Rockefeller Foundation of 6 million dollars for a 200-inch telescope. The money was given to the California Institute of Technology, on condition that a program of cooperation be established with the Mount Wilson Observatory and its owner, the Carnegie Institution of Washington.

Twenty years later, ten years after Hale's death, the telescope would be dedicated on Palomar Mountain in southern California. Since then it has continued to push back the frontiers of the universe as each of Hale's great telescopes had previously done. At the dedication in 1948 the 200-inch telescope towering above the distinguished audience was fittingly named the Hale telescope after the man "whose vision and leadership made it a reality." In 1970 the Mount Wilson and Palomar Observatories were renamed the Hale Observatories.

"It is perhaps symbolic of this man of great gifts and wide horizons," Walter Adams wrote, "that he who had devoted his life to the nearest star should find his last deepest interest in an instrument destined to meet the remotest objects of our physical universe."[30]

> It seems to me that through most of my life I have been fated to have lucky opportunities thrust before me and then to get undue credit for merely trying to carry them out, when as a matter of fact other and abler men did most of the work....
>
> ... The truth is, of course, that I have been enjoying from boyhood the things I liked most to do, and why should one be praised for simply having a good time? If this has helped other men to enjoy themselves also, this has added to my pleasure.[31]
>
> George Ellery Hale

99   The Hale Solar Laboratory and the 200-inch Telescope

Hale's library in the Hale Solar Laboratory.
Niels Bohr Library.

Walter Adams, who succeeded Hale as director of the Mount Wilson Observatory, in the doorway of the Solar Laboratory. The symbols above the door reflect Hale's long interest in the sun and in Egyptology.
Photo by Edison R. Hoge. Niels Bohr Library.

Albert Einstein during a visit to the Hale Solar Laboratory in the early 1930s.
California Institute of Technology Archives.

101 The Hale Solar Laboratory and the 200-inch Telescope

Dr. Wickliffe Rose
President International Education Board
61 Broadway New York

University Club
New York
April 16, 1928

Dear Doctor Rose—

In reply to your request I beg to summarize briefly the chief arguments favoring the construction of a looish telescope and to outline the procedure in view.

(1) The 100-inch Hooker telescope of the Mount Wilson Observatory has solved many fundamental astronomical and physical problems beyond the reach of our 60-inch reflector, and prepared the way for an attack on still more important problems which demand greater light-gathering power. Among these outstanding questions in their order are:

(a) The structure of the universe, calling for a more intensive study of the Galaxy, of which our solar system is a minute part, and especially of the vast region of spiral nebulae (island universes) beyond the Milky Way, where the 100-inch telescope has fixed the distance of the two nearest spiral nebulae at about one million light-years

and disclosed their true nature by partially resolving them into stars. It has also revealed hundreds of thousands of more remote spirals, many of which could be resolved, analyzed and measured with a larger instrument.

(b) The evolution of spiral nebulae, partially suggested by our recent studies.

(c) The evolution of stars, showing their origin, sequence, and physical and chemical development throughout their life history.

(d) The constitution of matter, since the enormously greater range in mass, temperature, pressure, and density of the heavenly bodies present opportunities for discovery far beyond the possibilities of laboratory experiments.

Scores of other problems calling for a larger telescope might be mentioned if space permitted.

(2) No method of advancing science is so productive as the development of new and more powerful instruments and methods. A larger telescope would

not only furnish the necessary gain in light, space-penetration, and photographic resolving power, but permit the application of many new ideas and devices derived chiefly from the recent fundamental advances in physics and chemistry. These advances, which have suddenly transformed spectroscopy from an empirical into an exact and rational science, would undoubtedly render possible many new and extraordinary discoveries with such an instrument.

(3) The time is also especially opportune because of recent engineering and optical progress, such as the development of fused quartz as a very advantageous substitute for glass, and the proof now available that the atmospheric conditions on Mount Wilson are sufficiently perfect to permit a large increase in aperture to be fully utilized. Hitherto, in the absence of such proof, increases in the size of telescopes have been made at the risk of defeat by atmospheric disturbances.

(4) A study of the optical possibilities, which depend chiefly upon the promise of success in the manufacture of a large quartz disc for the paraboloidal telescope mirror, convinces us that the diameter of this disc should be 200 inches. This would collect nearly four times as much light as the 100-inch telescope (the largest yet constructed), and reveal hundreds of millions of stars beyond its range, penetrate twice as far into space. The shorter relative focal length adopted in our design and the probable substitution of quartz for glass would greatly increase this gain, which could be still further enhanced by the improvement of photographic plates and accessory apparatus.

(5) The procedure in view involves: Insert(a) over

(A) an experimental test of the possibilities of fused quartz, and the manufacture as soon as feasible of a 200-inch mirror disc. I have already outlined to you the steps in this process, which would be carried on under the personal supervision of Dr. Elihu Thomson of the General Electric Co.,

Draft of Hale's letter to Wickliffe Rose advancing the idea of a 200-inch telescope, April 16, 1928. *The Hale Papers.*

Glass for the 200-inch disc being poured into its mold inside a beehive at the Corning Glass Works. *Corning Glass Works.*

The 55-foot tube for the telescope. Albert Einstein was among the scientific and industrial leaders who attended the ceremonial fitting of the last bolt into the skeletal mounting in April 1937.
Westinghouse Electric Corporation.

A portion of the glass disc showing the honeycomb pattern of the ribbed side. The lens of the Yerkes telescope could fit into the central hole of the disc being measured here by George V. McCauley of Corning.
Corning Glass Works.

Horseshoe bearing for the 200-inch telescope mounting on board a freighter bound for San Diego via the Panama Canal during the summer of 1938.
Westinghouse Electric Corporation.

The 200-inch mirror en route to Palomar Mountain in November 1947. Since April 1936 it had been at the California Institute of Technology for grinding and polishing.
Los Angeles Times.

The Mount Wilson — Palomar area. Map by Milford Zornes.
Niels Bohr Library.

107   The Hale Solar Laboratory and the 200-inch Telescope

The Pasadena Star-News, June 4, 1948.

Mt. Palomar Telescope Named in Honor of George Ellery Hale

200-INCH SKY-CAMERA DEDICATED

Astronomer Who Had Idea for Giant Eye Paid Tribute

By ROBERT CAHN

PALOMAR OBSERVATORY, June 3. (Special to The Star-News) —One of the world's greatest scientific achievements officially came to life here today as the 200-inch telescope was dedicated formally and received a new name—The Hale Telescope.

Before a large group of scientists, newspapermen and photographers, California Institute of Technology's eminent president, Dr. Lee A. DuBridge, this afternoon announced in a dedicatory address that the world's largest telescope would bear the name of a Pasadenan, the late Dr. George Ellery Hale, for many years director of Mt. Wilson Observatory, who originated the idea for the 200-inch and made possible its design and construction.

"As this great instrument probes the secrets of the universe, it is fitting that it should stand also in memory of the great scientist and the great leader who contributed so brilliantly to the science of astronomy and who served so able his community and his nation," Dr. DuBridge stated.

Dr. Hale's Widow Present

Dr. Hale's widow, who lives in Pasadena, was present at today's ceremonies and also witnessed the establishment of a bronze plaque in honor of Dr. Hale, which has been placed below his bust in the foyer of the observatory building which houses the telescope.

Scientists from all parts of the nation talked with the press, listened to the dedicatory speeches and then watched a demonstration of the telescope in action.

Further tribute to Dr. Hale and also to Andrew Carnegie were given by Dr. Vannevar Bush, president of Carnegie Institution of Washington, which will share in the operation of the observatory through its Mt. Wilson Observatory staff. These two observatories constitute the greatest concentration of astronomical power in the world, Dr. Bush said.

Speaking on "The Challenge of Knowledge," the Rockefeller Foundation president, Dr. Raymond B. Fosdick, sounded a warning in saying that there may be danger in the knowledge that will come from this mighty instrument.

Insanity of War

"This telescope may conceivably give us knowledge which, if we so choose, we can employ in the insanity of a final war," Dr. Fosdick said in comparing it with the giant cyclotron built at the University of California as an object for pure research, but later used for the atomic bomb.

Our knowledge must be anchored to moral foundations, he said, and added that the world also needed the perspective of the astronomer and the sense of proportion which this telescope can bring to mankind.

"In the last analysis, the mind which encompasses the universe is more marvelous than the universe which encompasses the mind," Dr. Fosdick concluded.

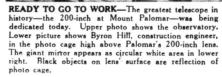

READY TO GO TO WORK—The greatest telescope in history—the 200-inch at Mount Palomar—was being dedicated today. Upper photo shows the observatory. Lower picture shows Byron Hill, construction engineer, in the photo cage high above Palomar's 200-inch lens. The giant mirror appears as circular white area in lower right. Black objects on lens' surface are reflection of photo cage.

HIS IDEA—The late Dr. George Ellery Hale conceived the idea of a new and more powerful telescope to probe into the mysteries of the uncharted skies. Tribute was paid Dr. Hale at today's dedication.

Participants at the ceremonies.
Los Angeles Times photo by Gordon Wallace.

Ira Bowen, Walter Adams, Lee Du-Bridge, and Hale's widow Evelina at the dedication ceremonies, June 3, 1948.
Niels Bohr Library.

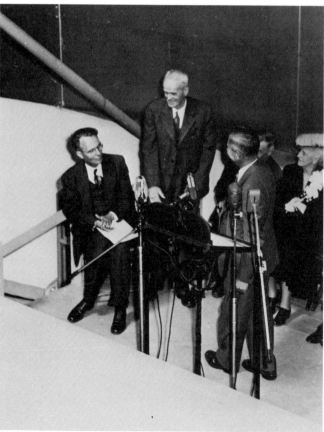

Notes

1. Hale to Harry M. Goodwin, December 13, 1932. The original letter is in the Henry E. Huntington Library and Art Gallery, San Marino, California, HM no. 28541.

2. Hale, "Biographical Notes," typescript, unpublished, pp. 4 and 6. The typescript is a part of the Hale Papers in the Hale Observatories Library and in the California Institute of Technology Archives (hereafter cited as the Hale Papers.)

3. Hale to Harry M. Goodwin, August 4, 1889. The original letter is in the Henry E. Huntington Library and Art Gallery, San Marino, California, HM no. 28412.

4. Reminiscences of Hale's first visit to Lick Observatory, written in honor of Ambrose Swasey. The typescript, 1933, is in the Hale Papers.

5. Hale, "Beginnings of the Yerkes Observatory." Typescript, 1922. Address delivered *in absentia* at the Yerkes meeting of the American Astronomical Society. The typescript is in the Hale Papers.

6. Ibid.

7. Hale to James Hall, July 12, 1895. Copy of the letter is in the Yerkes Observatory Records, Williams Bay, Wisconsin.

8. Hale to Simon Newcomb, January 5, 1899. The original letter is in the Library of Congress, Manuscripts Division.

9. Hale to William Rainey Harper, September 19, 1898. The original letter is in the University of Chicago Archives.

10. Trust Deed, *Carnegie Institution of Washington Year Book, No. 1, 1902* (Washington, D.C., 1903), p. XIII.

11. Hale to Mary Vaux (Mrs. Charles D.) Walcott, July 16, 1929. The original letter is in the National Academy of Sciences Archives.

12. Walter S. Adams, "Early Days at Mount Wilson," *Publications of the Astronomical Society of the Pacific, 59,* 213, 1947.

13. Robert S. Woodward to Hale, July 29, 1908. The original letter is in the Hale Papers.

14. Hale. *The Study of Stellar Evolution.* The Decennial Publications, second series. Vol. X (Chicago: The University of Chicago Press, 1908).

15. Hale to John D. Hooker, July 27, 1906. Carbon copy of the letter is in the Hale Papers.

16. Allan Sandage, "Galaxies," p. 4. In *The Hubble Atlas of Galaxies* (Washington, D.C.: Carnegie Institution of Washington, 1961).

17. Hunter Dupree, Review of *Explorer of the Universe,* by Helen Wright. *American Historical Review, 72,* 1115, April 1967.

18. Hale to Charles D. Walcott, January 25, 1908. The original letter is in the National Academy of Sciences Archives.

19. Hale to Charles D. Walcott, May 17, 1912, p. 3. The original letter is in the National Academy of Sciences Archives.

20. Hale, "War Services of George Ellery Hale." Unpublished manuscript, undated. The original is in the Hale Papers.

21. Hale to James A. B. Scherer, May 9, 1908. Carbon copy of the letter is in the Hale Papers.

22. Ibid.

23. Hale to Sir Ernest Rutherford, February 2, 1920. Carbon copy of the letter is in the Hale Papers.

24. Hale to Henry E. Huntington, May 11, 1914. The original draft letter is in the Hale Papers.

25. Hale, "Visual Observations of the Solar Atmosphere," *Proceedings of the National Academy of Sciences, 12,* 286-295, 1926.

26. Hale to his nephew, George Ellery Hale, March 30, 1926. Copy of the letter is in the Hale Papers.

27. Hale to Harry M. Goodwin, May 15, 1933. The original letter is in the Henry E. Huntington Library and Art Gallery, San Marino, California, HM no. 285430.

28. Hale, "The Possibilities of Large Telescopes," *Harper's Magazine, 156,* 639-646, April 1928.

29. Hale to Wickliffe Rose, February 14, 1928. The original draft letter is in the Hale Papers.

30. Walter S. Adams, "George Ellery Hale," *Astrophysical Journal, 87,* 369-388, 1938.

31. Excerpts from a letter from Hale to Harry Goodwin, April 7, 1936. The Henry E. Huntington Library and Art Gallery, HM no. 28548.

George E. Hale observing with the spectrograph of the Snow telescope.
The Hale Observatories.

2
Selected Papers of
George Ellery Hale

Introduction

Hale wrote and published over 450 articles and monographs, including scientific reports of his research, popular reviews of recent work in astronomy and astrophysics, and programmatic essays arguing for changes in scientific education and scientific organizations. The five presented here are representative of all of these categories and were selected because of their special significance in Hale's work.

His M.I.T. physics thesis (1890), written in longhand at the age of 21, demonstrated that he was already a professional with a perceptive overview of the past development and current state of his chosen field of solar studies. The style and content reveal his attention to the meshing of details and his technical mastery of the subject. His straightforward, first-person account is at the same time confident, critical, and respectful. His graphic account of his invention and use of the spectroheliograph delineates the enormous difficulties that had to be overcome before successful photographs of solar prominences could be obtained and the future use of this extremely valuable tool could be assured.

In "A Plea for the Imaginative Element in Technical Education" (1907), Hale urged his former M.I.T. teachers to broaden scientific and technical education by including more humanistic elements. He called for an approach that would emphasize the unity of science, reflecting his own efforts to bring chemistry, astronomy, and physics closer together to study the processes of stellar evolution. The evolution of stars could be related to the history of the evolution of the earth and of man and his social behavior, thus linking science and the humanities. In this essay, Hale identifies many of the problems that are still plaguing science educators.

In his 1909 address on "Solar Vortices and Magnetic Fields," Hale described for his Royal Institution audience the research results that had been obtained with the new instruments, institutions, and research teams he had invented and organized. He explained the methods that made possible his discovery of the magnetic fields in sunspots. These methods were guided by the same philosophy that led to his creation of large observatories as experimental laboratories where the nature of celestial objects could be analyzed with powerful instruments.

Hale was deeply concerned with the appropriate role of scientific organizations, and his essay on "National Academies and the Progress of Research" (1914) stressed that the Academy had to adapt to new national and scientific needs of the day and to be prepared for the future. He called for changes in the internal organization and external role of the Academy. Here again, he emphasized the importance of the study of stellar, human, and social evolution as a way of improving the understanding of science.

Hale was developing compelling and successful arguments for funds to construct a 200-inch telescope when his 100-inch, the world's largest, had been in use for less than a decade. In his 1928 article on "The Possibilities of Large Telescopes," he outlined the dramatic results obtained with the 100-inch, the new questions it had opened up, and the desirability and feasibility of constructing a still bigger telescope to find the answers.

Prominence observed March 14, 1869, 11 h. 5 m. by Sir Norman Lockyer.
J. Norman Lockyer, The Spectroscope and its Applications (London: Macmillan, 1873).

Photography of the Solar Prominences

The research described in the following pages claims to be nothing more than a beginning in an extended series of solar investigations which I hope soon to continue.

Introduction

No branch of physics seems to me to offer richer returns to the patient investigator than that dealing with the constitution of the sun.

Although the work has been in progress for more than twenty years, questions of every degree of complexity yet remain to be answered. We are still compelled to wait for the rare and fleeting visits of the lunar shadow to show us the corona. And even when seen, the remarkable green line in its spectrum remains as much of a mystery as ever, and no terrestrial element has been found with any line corresponding to it in wave-length.

The D_3 line is equally peculiar in having its origin unknown, and many lines seen in the spectra of sun-spots might be placed in the same category. But space forbids the mention of even a tithe of these interesting instances, for they seem well nigh innumerable.

The prominences have been perhaps as thoroughly studied as any of the solar phenomena, but they still have many secrets to reveal.

It has long been my conviction that photography might well play a prominent part in further attacks.

With this end in view I have carefully considered the various difficulties to be overcome, and thus have been led to devise the simple methods outlined in this paper.

Early Eclipses

It is a remarkable fact that no observation of the chromosphere or prominences was recorded prior to 1706. At the eclipse of that year Captain Stannyan, observing at Berne, noticed upon the western limb of the sun a blood red streak of light which was visible for several seconds. In 1715 Halley and Louville described a similar phenomenon. But it was not until 1733 that the prominences were seen by Vassenius in the form of small pinkish clouds, floating above the moon's limb in what he supposed to be a lunar atmosphere.

In succeeding eclipses up to 1842, it is doubtful whether the chromosphere and prominences were seen at all, as no definite mention of them was made. But at the eclipse of July 1842, their great brilliancy attracted the attention of many skilled astronomers. A vigorous discussion ensued as to their real nature. Some held with Vassenius that the prominences were lunar clouds, others, including Arago, believed

Thesis for the B.S. in physics (M.I.T., 1890).

them to be clouds in the sun's atmosphere. Some thought them to be solar mountains, others maintained that they were gigantic flames, and there were others who considered them nothing more than optical illusions. But the eclipse of 1851 strengthened the evidence of their solar origin, and a theory, substantially that enunciated earlier by Arago, became quite generally accepted. Their real existence in the solar atmosphere was finally established by the photographs of De La Rue and Secchi at the eclipse of 1860.

Meanwhile the spectroscope had demonstrated its analytical power, yielding in the hands of Bunsen and Kirchhoff a means of determining the chemical constitution of the sun. At the eclipse of 1868 spectroscopes were attached to the telescopes of many observers, and it was at once concluded that the "lunar clouds" of Vassenius were for the most part vast masses of hydrogen gas rendered incandescent by the intense heat of the sun. This conclusion has not lacked abundant confirmation.

Method of Observation

Up to this time these phenomena had only been seen at the rare occurrence of a total eclipse, and it was not surprising that such limited and infrequent observations had discovered but little of their true nature. It was eminently desirable, therefore, that some method should be devised by which they could be observed at any time in full sunshine. The problem was, to so reduce the light of the atmosphere that the prominences should no longer be hidden by its glare.

The first attempts were made by Mr. (now Sir William) Grove, who made use of red glass in observing the limb of the sun, the telescopic image being stopped out by a diaphragm. But he was unsuccessful, for the atmospheric light was still sufficiently intense to completely drown the image of the prominence.

In October, 1866, Mr. Norman Lockyer suggested that a spectroscope be used to overcome the difficulty.[1]*

The atmospheric light gives a continuous spectrum, crossed of course by the dark Fraunhofer lines. The brilliancy of such a spectrum can be decreased any desired amount by multiplying the number of prisms in the spectroscope. On the other hand as the prominences are composed largely of hydrogen, most of their light is concentrated in the four or five lines of the hydrogen spectrum and increase of dispersive power only serves to separate these lines more widely, without materially affecting their brilliancy. It is thus seen that there is

*Editors' note: All footnotes in the thesis were originally numbered on each page, but here they have been numbered continuously and the numbers moved to the end of sentences and listed on pages 138-139. Hale's citation style has been preserved, but some supplementary information has been added to the notes.

a limiting dispersion, at which the bright lines of hydrogen are of the same brightness as the overlying continuous spectrum of the atmosphere. With a somewhat increased dispersion the bright prominence lines should be distinctly seen upon a less brilliant background of the same color, when the image of a prominence is brought over the slit of a powerful spectroscope.

Acting on this idea, Mr. Lockyer had constructed a spectroscope provided with seven 45° prisms of dense flint glass. It was attached to an equatorial refractor of $6\frac{3}{4}$ inches aperture and about 8 ft. focal length; the image of the sun on the slit plate was therefore something less than 1 inch in diameter. The completion of the spectroscope was so much delayed, that it was not tried until October 16, 1868.

The search for bright lines was for a time unsuccessful, but on October 20 Mr. Lockyer announced to the Royal Society the existence of three bright lines in the following positions —

"   I   Absolutely coincident with C.
    II Nearly coincident with F.
    III Near D.

The third line is more refrangible than the more refrangible of the two darkest lines by eight or nine degrees of Kirchhoff's scale."[2]

Curiously enough, the letter announcing this discovery to the French Academy of Sciences was followed the same day by a letter from J. Janssen who was just about to return from observing the solar eclipse at Guntoor.

During totality he had been so struck with the brilliancy of the bright lines seen through his spectroscope, that he attempted the next day to see them again, and at once succeeded. He was thus enabled to continue his observations from August 19 until September 4, "a period" to use his own words, "like an eclipse of seventeen days."

Thus established, the new method at once came into general use. But as yet only a rough idea of the true form of the prominences could be made out by moving the solar image across the slit. What was desired was a means of seeing the whole prominence at once, and not alone a small strip through a narrow slit. At the very outset Janssen and Lockyer tried to form persistent images on the retina; the former by giving a rotatory motion to a direct vision spectroscope, the latter by causing the slit to rapidly oscillate. Prof. Young also used an oscillating slit, placing at the focus of the eye-piece a diaphragm which should move with the slit, thus cutting off the light of the neighboring portions of the spectrum. But he found that "although seen in this way, the prominences appear very bright, yet the working of the apparatus always causes a slight oscillation of the equatorial, which interferes with the definition of details."[3] He retained, however, the moveable slit, finding it very convenient in observations of spots or prominences.

The method now ordinarily employed was first proposed by Zöllner in a paper communicated to the Royal Saxon Academy of Sciences in February, 1869. He argued that the brightness would be much diminished by using a moveable slit, and ended with the statement that "it is only necessary to open the slit so far that the protuberance or a portion of it, appears in the opening. By polarizing or absorbing media, placed before the eye-piece, the light in the whole field of view can be so diminished that the proper relation of intensity between the protuberance and the superposed spectrum may be obtained."[4]

A few days later Mr. Huggins published in the Proceedings of the Royal Society of London, an account of a successful observation of a prominence through an open slit, the stray light being reduced by the use of ruby glass and a diaphragm placed at the focus of a positive eye-piece.[5]

It was soon seen by Lockyer that the ruby glass used by Huggins was unnecessary, and that it was only essential to open widely the slit of the spectroscope. The width to which it may be opened depends upon the dispersion of the instrument employed, and also upon the condition of the atmosphere. If the sky has a whitish appearance, due to the presence of ice particles, a narrow slit must be used, if observation can be carried out at all.

Another method which has been suggested for observing the prominences, necessitates the use of a direct vision prism, placed before the slit, or a prism of small angle, placed just outside the object glass of the telescope. The dispersion could be thus very largely increased. Such an arrangement is very highly recommended by Secchi, as it allows not only the chromosphere and prominences to be observed, but also the photosphere and spots at the same time.[6] This is very desirable at times, as the relation of a prominence to a neighboring spot would be at once recognized.

In 1872, Messrs. Lockyer and Seabroke, described to the Royal Society, a method of observing the whole of the chromosphere by means of a ring slit as follows. "The image of the sun is brought to focus on a diaphragm having a circular disc of brass (in the centre) of the same size as the sun's image, so that the sun's light is obstructed and the chromospheric light allowed to pass. The chromosphere is afterwards brought to a focus again at the position usually occupied by the slit of the spectroscope; and in the eye-piece is seen the chromosphere in circles corresponding to the C of other lines."[7] The same device had also been tried by Zöllner and Winlock, but without success. I shall again refer to it in this paper.

Prominence Photography

As photography has been so largely employed in almost every department of astronomical research, and with such remarkable success, it is rather to be wondered at that very little has been done toward photographing the prominences.

A simple and reliable method would not only do away with the tedious and inaccurate task of drawing, but might possibly discover new phenomena of value in clearing up the perplexing questions of the solar theory. A series of photographs of the same prominence, taken at exactly equal intervals of time, would be of much greater value than a similar series of drawings. And if in time the method could be perfected and made automatic, so that it would photograph the whole circumference of the limb at equal intervals throughout the day, its value would be beyond question.

There are yet other opportunities for photography. In eye observations we are limited to the visible portion of the spectrum; photography would lay open the ultra violet and ultra red.

It would permanently register all the phenomena of the distortion and displacement of lines in the spectrum, and render their measurement a leisurely and exact process with the dividing engine. And it might ultimately lead to a more certain knowledge than we now possess of the *true* form of a prominence.

In an observation of a prominence through, say, the C line, what is seen is that portion of the prominence which is composed of incandescent hydrogen. A photograph of the prominence taken through an iron line, if one were visible at the time, would show the region occupied by iron. In the same way the parts played by calcium, sodium and other elements might be shown; and a composite photograph from such a series of negatives would for the first time (outside of an eclipse) show the true form. It is now known from observations made just before and after eclipses, that the C line gives the true form of the prominences very nearly if not exactly, but there is still room for plenty of work in this direction and in many others. For instance, is the form seen through D_3 *exactly* the same as that seen through C? If not, are they more alike at some times than at others? Answers to questions such as these might render less mysterious the true nature of "helium," and define its relation to hydrogen.

Perhaps sufficiently good photographs for these purposes will never be made;—certainly they would surpass anything we could ever hope to get with a horizontal telescope in the atmosphere of Cambridge, but it seems to me that in the lapse of time some of our mountain observatories will accomplish all this and much more.

Account of Previous Methods

The first experiments in prominence photography were made by Prof. Young in 1870. The following is his own account of the work:

"The protuberances are so well seen through the F and 2796 (K) lines that it is even possible to photograph them, though perhaps not satisfactorily with so small a telescope as the one at my command. Some experiments I have recently made show that the time of exposure, with ordinary portrait collodion, must be nearly four minutes, in order to produce images of a size which would correspond to a picture of the solar disc about two inches in diameter.... Negatives have been made which show clearly the presence and general form of protuberances, but the definition of details is unsatisfactory....

"We worked through the hydrogen γ line, which though very faint to the eye, was found to be decidedly superior to F in actinic power. The photographic apparatus employed consisted merely of a wooden tube, about 6 inches long, attached at one end to the eye-piece of the spectroscope, and at the other carrying a light frame. In this frame was placed a small plate-holder, containing for a sensitive-plate an ordinary microscope slide 3 inches by 1."[8]

We see, then, that Prof. Young brought the γ hydrogen line into the field, opened the slit widely, and made it tangent to the sun's limb at a point where a prominence was known to be. An exposure of about four minutes was then made and this produced a negative of the prominence upon a wet collodion plate. Prof. Young has very kindly shown me silver prints from the best original negatives. In these, little more than the general outline can be seen. This is due partly to a slight displacement of the image during the exposure, as the polar axis of the telescope was slightly out of adjustment. But a much more radical defect lies in the broad and nebulous character of Hγ. As seen through this line every point in the prominence is drawn out into a short line at right angles to the slit, thus rendering good definition impossible.

Neglecting for the present the claims of Zenger I have been unable to find any account of other photographs of the prominences taken without an eclipse.[9] But methods have been invented for the purpose, and these merit description.

In 1874 Dr. Lohse made several unsuccessful attempts to photograph the chromosphere and prominences by direct methods — i.e. — using the direct image of the sun without a spectroscope. In 1880 he devised a special form of apparatus. It consisted of a direct-vision spectroscope held in a suitable frame, with its axis of collimation parallel to the axis of the telescope to which it was attached. The centre of the sun's image was brought on to the axis of the telescope, so that the slit was radial. A hand motion then served to rotate the spectroscope about the telescope's axis, and in this way it was hoped to build up an image of all the prominences on a stationary photographic plate at the focus of the spectroscope. Before the plate a slit was placed to cut out any desired region in the spectrum. The line Hγ was ordinarily used.[10]

Several objections might be raised against this instrument.
1. The dispersion of the direct-vision prism was necessarily small, and therefore the atmospheric light was not greatly reduced.
2. The nebulous line Hγ was used. The difficulty with this line has already been mentioned.
3. The motion of rotation was produced by hand, and was therefore more or less unsteady. Even with a good clock, my experience has shown the extreme difficulty, if not impossibility, of obtaining perfectly uniform motion; and of course hand motion is incomparably worse.

These combined defects lead me to doubt the value of Lohse's rotating spectroscope, and the fact that no photographs taken with it have been mentioned strengthens this opinion.

In the paper before quoted from, Lockyer and Seabroke propose to use their ring slit in photographing the prominences in the following manner: — "A large Steinheil spectroscope is used, its usual slit being replaced by the ring one.

A solar beam is thrown along the axis of the collimator by a heliostat, and the sun's image is brought to a focus on the ring slit by a $3\frac{3}{4}$ inch object-glass, the solar image being made to fit the slit by a suitable lens. By this method the image of the chromosphere received on the photographic plate can be obtained of a convenient size, as a telescope of any dimensions may be used for focusing the parallel beam which passes through the prisms on to the plate."[11]

Drawings were exhibited showing the prominences observed with the ring slit, but no photographs were shown, probably because none had been obtained. The apparent width of the slit upon the photographic plate would necessarily be quite large, if any prominences of even medium height were on the limb. Thus a large amount of atmospheric light would be admitted, which would seriously fog the plate unless very high dispersion were employed.

In 1879 a letter was published in the *Comptes Rendus* describing a method imagined by C. W. Zenger for photographing the chromosphere, prominences, and corona without the use of a spectroscope.¹²

The plate was first put into a solution of pyrogallic acid and citrate of silver, and then given a very short exposure to the direct solar image, using "une couche absorbant tous les rayons dont est composée la lumière de la couronne et des protubérances solaires." The author goes on to add:— "C'est en étudiant par le spectroscope des pellicules ainsi obtenues, que j'ai constaté l'absorption de raies caractéristiques de la couronne et des protubérances, et c'est pourquoi les protubérances et la chromosphère sur les épreuves négatives, apparaissent blanches, la couronne en est moins prononcée, seulement blanchâtre, ce qui montre que la lumière coronale est très distincte de celles de la chromosphère et des protubérances."

Altho a number of photographs, said by M. Zenger to show the prominences and corona, were forwarded to the Academy, I should be inclined to hesitate in pronouncing upon the value of the method, at least until the composition of the remarkable absorbing solution were made known.

Dr. Janssen has also tried to use the direct solar image. In a short note on the subject he says:

"Il faut que l'action lumineuse solaire s'exerce assez longtemps pour que l'image solaire devienne positive jusqu'aux bords, sans les dépasser. Alors la chromosphère se présente sous forme d'un cercle noir, dont l'épaisseur correspond à 8″ ou 10″.¹³

In this case, and in all others where a direct image of the sun is received upon a photographic plate, it is very improbable that chromosphere or prominences produce any appreciable effect. The "black circle" is solely due to the bright disc of the sun, and would be formed, even if the chromosphere did not exist.¹⁴

New Methods Proposed

I believe that the following considerations will ultimately lead to a successful solution of the problem:—

1. A sufficient dispersion may easily be obtained by the use of a large diffraction grating, mounted with a pair of telescopes of large aperture. This grating should be very bright in the second or third order, if a ruling with about 14,000 lines to the inch is used. Suitable absorbing substances will be required before the slit to cut out the overlapping spectra.

2. It is a well known fact that the solar line C is by far the sharpest line in the hydrogen spectrum. It is almost exclusively used in observations of the prominences because of the brilliant and well-defined

images seen through it. Therefore it is desirable that this line be selected in preference to any other for prominence photography. Hence special plates will be required, which are sensitive to this region of the spectrum.

D_3 is another very sharp line and as it is always visible as far above the limb as C, it is also recommended for photographic work, tho the extreme brilliancy of this part of the spectrum may prove troublesome. Plates sensitive to the yellow rays will of course be needed for this region.

3. Instead of moving the whole spectroscope as proposed by Lohse, I have devised the three following methods:

(a) Change the rate of the driving clock of the telescope, so as to make the sun's limb move slowly across the slit of the spectroscope. Move the photographic plate in the plane of dispersion and at right angles to the axis of the observing telescope; its velocity depending upon the ratio of the focal lengths of the collimating and observing object glasses; if these are equal, the velocity must equal that of the sun's image. Thus a series of images of the slit would be found, side by side, and merging into each other, thus building up an image of the prominence.

This method would allow the use of a narrow slit, thus greatly reducing the atmospheric light, and increasing the sharpness of the photograph. A second slit should be placed immediately in front of the plate, and so nearly closed as to allow only the C (or D_3) line to pass through. This would cut out the troublesome continuous spectrum on either side, and leave the photograph of the prominence in strong contrast on a clear background.

In practice it is impossible to fully realize this, but still the second slit can be closed to such an extent as to greatly reduce the superfluous light. The use of a second slit is in itself not new, as it was employed by Huggins and others as an aid in observation, and also, as we have seen, by Lohse.

It is easy to see that a radial slit would probably prove most useful for this work. But there are, at a given time, only two positions on the sun's limb at which the slit may be radial, when the direction of the sun's motion is at right angles to it. If the prominences did not happen to be at either of these points, they could not be photographed to the very best advantage. But it is not absolutely essential that the motion should be at right angles to the slit, and in this case a prominence could be photographed at any point on the limb. The chromosphere, and all the prominences visible, could be taken upon one plate, by reducing the diameter of the solar image until it became less than the length of the slit; the sun being then allowed to move across centrally.

The above arrangement may be modified in such a way as to greatly increase its value and extend its range of application. By a suitable combination of prisms or mirrors, cause the image of the sun to rotate about its center in such a way that the whole circumference of the limb will move across the center of a radial slit.

The clock of the equatorial used, should be electrically controlled, so as to keep the sun's center in a fixed position with regard to the slit. Arrange a brass disc at the focus of the spectroscope, and cause it to rotate synchronously with the solar image by means of a tangent screw connected with a clock, which also drives the device for rotating the sun. On the brass disc secure a photographic plate by spring clips, and let the distance from the center of rotation to the edge of the spectrum on the plate be equal to the semi-diameter of the sun's image, supposing that there is no magnification in the spectroscope. Then place a stationary second slit just in front of the plate and through it allow C (or D_3) to pass. It is thus seen that one complete revolution of the plate and solar image would photograph the whole chromosphere and any prominences present on the limb.

A simple addition would make the apparatus entirely automatic so that it might be set on the sun in the morning and allowed to run all day, making complete pictures at intervals equal to the time of one revolution. Substitute for the glass plate used above a photographic roll holder loaded with a thin sensitive film. Place a spiral spring, or cord and weight, on one of the rollers so that on releasing a catch the film will be transferred from one roll to the other. At the end of each revolution of the disc a platinum point should make contact with a light spring, thus throwing into circuit a small electro-magnet attached to the disc, and revolving with it. This would release the catch, and enough film for a new exposure would be brought into position.

(b) This method is similar in principle to the above, but the result is effected in a different way, making a radial slit applicable to any portion of the limb.

Provide the slit with a uniform motion across the axis of the collimator. Before the photographic plate arrange a second slit, and cause this to move at the same rate as the first slit, (if there is no magnification in the instrument) by means of a screw cut with the same thread; both screws being provided with grooved heads for a cord, and driven by a single clock. In this case the plate and sun's image must be stationary. The slit then moves across the prominences and the second slit prevents the plate from fogging.

This device seems to promise greater usefulness than any other proposed. It can, of course, only be used on an equatorial telescope provided with a good clock.

(c) This method was suggested by W. B. Hale and is a modification of Lockyer's ring slit. Mechanical difficulties would probably render its actual operation very difficult, but it is at least interesting theoretically.

Suppose the ordinary slit of a powerful spectroscope removed, and for it substitute a ring slit made in two parts; the outer like the "iris" diaphragm used on photographic cameras, and the inner a disc capable of radial expansion. Suppose the disc to be fixed within the opening of the diaphragm, and the two so connected by gearing that the edge of the disc and the inner circumference shall always be separated by a very small distance — say a tenth of a millimeter. A clock motion connected with a rotating collar might then expand the diaphragm and disc together, so that the annulus of light would start from the sun's limb and travel slowly outward, passing over the chromosphere and prominences, and admitting but little atmospheric light because of the narrow slit.

The photographic plate would of course be held stationary at the focus, and the sun's image would necessarily be maintained in a fixed central position.

In all of the above methods slow development of the plates is desirable, in order to obtain the greatest possible contrast.

Description of Apparatus

Through the kindness of Prof. E. C. Pickering, the 15 inch equatorial of The Harvard Observatory was placed at my disposal in November 1889.

But in the opinion of Mr. George Clark the weight of my large spectroscope was considered too great to be safely carried by the wooden tube of the equatorial, and it was finally decided to adapt the instrument to the 12 inch horizontal telescope, which Prof. Pickering very kindly gave up to my use, altho it was at that time employed in photometric work. [See Plate I, page 140.]

The Horizontal Telescope

The mirror is 18 inches in diameter and 4 inches thick, and its surface is said by the Clarks to have a radius of curvature of several miles. It rests upon an iron plate in a cradle, which can be rotated by a friction roller attached to a long rod leading into the eye-end of the telescope. This is the motion in altitude. The axis of the telescope is at right angles to the meridian, and a screw at the back of the mirror gives the motion in azimuth. This screw is driven by an endless cord in connection with a clock placed convenient to the hand of the observer, so that its rate can be easily regulated by screwing down the friction disc of a centrifugal governor. As the mounting is alt-azimuth, the image of the sun cannot be kept stationary by the clock except at the meridian passage, and this is one of the strongest objections to the instrument, of course unfitting it for part of the work proposed. [See Plate II, page 141.]

The object glass is an excellent one, and was made by the Clarks. Its clear aperture is 12 inches and it has a focal length of about 1.7 feet. The image of the sun at the focus is therefore about 1.9 inches in diameter.

The Spectroscope

The spectroscope and eye end of the telescope are shown in Plate III [page 142].

The spectroscope was specially constructed for me by Mr. J. A. Brashear, of Allegheny. It has proved to be a most excellent instrument, and I am glad to testify to the fine workmanship of its skillful maker.

The slit is made of glass hardened steel, gilded to prevent rust, and opens equally in both directions from the center, the width being read off from a graduated head. The whole slit-plate is provided with a screw motion across the end of the collimator, a very convenient adjustment in prominence or spot observations and also necessary in the second method proposed for photographing the prominences. A simple little device of my own allows three spectra to be taken edge to edge on the same plate. It consists merely of three small strips of brass, each cut with a small window one-third as long as the slit. With the first one slid into position behind the slit the lower third is uncovered. Replacing this by the second uncovers the middle third, and the last one leaves all but the top third covered. [See Plate IV, page 142.]

The lenses of the collimator and observing telescope are exactly alike, $3\frac{1}{4}$ inches in diameter and $42\frac{1}{2}$ inches focus. Thus there is no magnification in the instrument when no eye-piece is employed, and the motion of the photographic plate will be the same as that of the sun in the first method proposed. The whole collimator tube can be moved by a screw through collars in the frame, and thus the slit can be brought exactly to the sun's focus.

The grating is one of the excellent rulings of Prof. Rowland, and is remarkably bright in one of the second spectra. It is ruled with 14438 lines to the inch on a highly polished surface of speculum metal, and contains altogether more than 48000 lines. It is mounted in a holder with adjusting screws at the top and side, and stray light is excluded by a cylindrical brass cover, provided with close fitting openings for the two telescopes. The circular plate upon which the grating holder stands is divided on its edge to degrees for convenience in setting, and a rod connected with a tangent screw carries the slow motion to the eye-end of the spectroscope. A quick motion is also supplied. Eye-pieces of various powers are used for observation, either with an adapter at the end of the tube, or at the side with a total reflecting prism when the sliding plate-holder is in position.

The sliding plate-holder was designed by myself for the first method of prominence photography. An ordinary $3\frac{1}{4} \times 4\frac{1}{4}$ plate-holder is held by a spring clip in a light frame made of brass tubing, which slides with little friction between V shaped guides. Its direction of motion is at right angles to the lines in the spectrum. The method of producing this motion will be described with the adjustments of the instruments. When in use the plate-holder, guides, etc. are completely enclosed by a tight fitting mahogany cover. To draw the slide, a brass rod is pushed in through a small opening on the left, and screwed to a threaded piece on the end of the slide.

Directly in front of the plate is an adjustable slit, the purpose of which has already been mentioned.

The whole spectroscope is supported in a strongly braced frame of steel tubing, and by means of a gear and either one of two pinions, it is easily rotated about the axis of the collimator. This allows the slit to be made tangent or radial at any point on the sun's limb.

Adjustments of Spectroscope

The spectroscope was first put in place on the horizontal telescope, and fixed in such a position that the sun's image was in focus on the slit.

The collimator and observing telescope were next roughly focused for parallel rays. The plane of the grating was then put at right angles to the plane through the axis of the two telescopes by placing a hair across the center of the slit, and bringing it on to a micrometer line in the center of the field by turning the back screw of the grating holder. The observing telescope was next removed from the spectroscope and carefully focused for parallel rays. The mean of a series of observations was taken, and the telescope replaced, and clamped at this reading. The grating was then turned until the reflected image of the slit came into the field. Then the focus of the collimator was adjusted until the edges of the slit in the image were perfectly sharp.

The draw-tube of the observing telescope was then unclamped, the photographic plate-holder put in position and secured from sliding, and a series of photographs taken of the third spectrum near G, the focus of the telescope being changed and recorded for each exposure.

From this series the sharpest was selected as most nearly giving the true focus for the part of the spectrum used, and the observing telescope was then clamped at this point in preparation for a series of exposures to find in what position of the grating its lines were parallel to the slit.

In order to record the various positions of the grating, measurements were made between each exposure from the plate on which the grating stands to the projecting arm of the grating holder. These were expressed in millimeters plus fractions of a turn of the adjusting screw.

In all, over 40 exposures were made in the adjustment of the spectroscope, G and the H and K lines of the third order being used.

Solar Observations

The spectroscope was in good adjustment and ready for solar work on Jan. 14, 1890. On that date the limb of the sun was brought to focus on the slit jaws by making the slit radial, and moving the whole collimator by the slow motion until the edge of the spectrum appeared sharp. Many attempts to observe the reversal of the C line were made with both radial and tangential slit at various points on the limb, but they were entirely without success. On the next date of observation a slight arrowhead appearance of C was suspected, but the definition of the sun's image on the slit was very poor, and work had to be suspended for this cause. In short, on the six available days for observation before Feb. 19, no decided evidence of reversal was obtained. On that date, by the aid of a piece of red glass before the slit, the arrowhead appearance was well seen, but nothing further could be made out. On Feb. 22, the general form of the chromosphere was seen, and D_3 observed for the first time.

These observations are mentioned here in order to give an idea of the way in which the definition varied, now allowing the general outline of the chromosphere to be discerned, and again rendering any work out of the question.

From this time until the middle of March there were but five clear days for observation, and on these the definition was variable, and the chromosphere but poorly defined, while no prominences had been seen. It was at length concluded that the trouble must be due in some way to the horizontal telescope, and on March 22 I carefully resilvered and polished the 18 inch mirror of that instrument. This to a great extent removed the difficulty, and on March 24, a small prominence was faintly observed. But still the mirror performed badly at times. On the morning of March 31 the mirror was set on the sun, and in a short time a small prominence was found. With the slit tangential, and not too wide, the definition was remarkably good, both in the prominence and in the chromosphere at its base. After observing at this point for a time I began a search around the limb, and soon noticed that the definition was changing rapidly, and becoming much poorer. I was so much struck with this fact, that I rotated the spectroscope into its first position, and attempted to observe the prominence so clearly seen before, but it was no longer visible, and the chromosphere was so blurred and indistinct that its form could not be made out. That this was not due to a change in the prominence itself, the poor definition at all other points on the limb attested. It was evidently caused by a change in the definition of the solar image given by the horizontal telescope, and this change seemed in all probability to be due to a distortion of the mirror by the sun's heat.

Some other observations led to the same conclusion. The slit was opened widely, and a portion of the limb made to lap over it. The grating was then rotated until it reflected the solar light directly, and the image was examined with a low eye-piece and shade glass. The limb was seen to be very poorly defined, and no changes in the focus on the slit or the focus of the eye-piece would make it sharp. A sun spot which happened to be on the disc at the time was examined in the same way with similar results.

On April 1 a small prominence was found in the same position as on the previous day. Observation was begun about 1^h30^m, and from that time until 3^h30^m, (when the sun was too far west for further work) the definition was excellent, in spite of the experience of the day before. Using the longest Beck eye-piece, and observing the image by direct reflection from the grating, it was found to be very sharply defined, and the mottling of the surface could be well seen.

The explanation of this variable definition is probably to be found in the condition of the atmosphere. On March 31 the air was warm and still, while the next day a brisk breeze was blowing, and it was much colder. This would tend to keep the density of the air between the mirror and object glass more uniform, and also keep down the temperature of the mirror, and diminish the effect of radiation. Not long after I paid a visit to Mr. George Clark at his shop, and found that he entirely agreed with me as to the cause of the differences in definition. His experience with the mirror when it was first set up is well worth noting. It had been tested at the shop and found to be perfect, after which it was taken to the Observatory, and placed in position. It was immediately found to be so concave that no star images of any sharpness at all could be obtained. When taken back to the shop the sodium flame at once demonstrated its perfect flatness. The mirror was then silvered, as the light in the first instance had been very weak. The Newton's rings test still showed it to be flat, but when mounted upon the iron plate of its stand it became decidedly convex, and again no sharp star images could be obtained. The difficulty was soon conjectured. The brick pier, having absorbed heat during the day, radiated it at night upon the back of the mirror. This caused it when first used to become concave. After the coat of silver had been put on, the front surface became a poor radiator of heat, and now, instead of being concave, the mirror was convex.

A partial remedy was found in covering the back and circumference of the mirror with bright sheet tin. This equalized the radiation, and secured very fair definition. But it is easily seen that exposure to the direct rays of the sun would very probably affect the definition by unequal heating, and this I have found to be the case. Perhaps the best way to obviate this would be to support the mirror only by its circumference, leaving the back surface open and silvered or not according to the condition of the front surface. It was found necessary to support in this way the small mirrors used on the Transit of Venus photoheliographs. But a mirror 18 inches in diameter could not be held in this way without bending, and thus destroying the definition by its own weight.

It has been said that the definition was good on cool, breezy days. As such days were not to be had for the asking, all that could be done was to make the best of as many as good fortune might bring. I was forced to this rather unsatisfactory conclusion, as Mr. Clark did not think it safe to attach the spectroscope to the 15 inch equatorial, and the 11 inch equatorial was so encumbered with heavy prisms and balancing as to render its use entirely impracticable in the short time remaining for my work. However, most of the time the definition has been fair, and by no means so hurtful to these photographic experiments as is the large amount of diffuse light.

My experience with the horizontal telescope does not lead me to echo the unqualified praise given by M. Thollon to this instrument.[15] It is true that in many respects it is remarkably well adapted to spectroscopic work, especially when a large and heavy spectroscope is used. The tube always remaining in the same position gives great rigidity, and the spectroscope is always conveniently situated for observation. In the case of the equatorial, the contrary is true. But an alt-azimuth mounting is very objectionable, and the range of a large mirror held in this way is exceedingly small. For instance, the instrument I have been using can be employed for solar work only from about 9^h30^m in the morning until about 3^h0^m in the afternoon, when the sun is not far from the equator. But as it was built for work near the meridian, a slight change in design would somewhat extend its range. I think it would also be much better for work on the sun if its mirror were much smaller, and supported only by its circumference, as recommended above. A still further improvement would be the use of some form of equatorial mounting, tho I doubt if a perfectly satisfactory siderostat has yet been constructed.

But supposing all these improvements added, there are yet difficulties from which the equatorial is free. Between the mirror and object-glass there is necessarily a considerable air-space, and here unequal heating must seriously affect the definition. And there is moreover a large amount of diffuse light reflected by the mirror—a most radical defect in prominence observation, and especially objectionable in such photographic experiments as I have undertaken.

This diffuse light will perhaps account for the extreme difficulty I have experienced in seeing F bright with a radial slit. With a tangential slit there has been less trouble, but observations have not been satisfactory.

The Method (a) in Practice

The various imperfections of the horizontal telescope already enumerated soon made it evident that no photographs of any intrinsic value could be obtained, but the possibility of demonstrating the practicability of the method still remained. For this purpose the F and h lines were largely employed, principally because of the difficulty in obtaining plates sufficiently sensitive for use with C and D_3. A discussion of the experiments with several organic dyes and other sensitizers for the less refrangible rays will be taken up later.

It has been said, that observations of reversal in the case of F and h were quite rare. This statement needs some qualification, but a word must first be said with regard to the magnitude and brilliancy of the prominences experimented upon. Only one prominence has been seen which could be called bright, and this was of small size, and in the most inconvenient position possible upon the limb. All others have been faint and cloud-like, tho they frequently have attained greater elevations. I do not consider this faintness to reside so much in the absolute brilliancy of the prominences themselves, as in the conditions under which they have been observed. In fact, from observations made with the 6 inch equatorial at Princeton—for the use of which I am much indebted to Prof. Young—I am convinced that the diffuse light from the mirror of the horizontal telescope must increase in a marked degree the difficulty of distinguishing the prominences.

This is no doubt the reason why good reversals in the blue and green have been so hard to obtain. With Prof. Young's refractor it was very easy to see the form of the prominence through F, while with my instrument only the merest outline can be seen even under the very best conditions, and aided with a screen of blue glass close in front of a tangential slit. With a radial slit I have seen F bright on only one occasion, and then it was extremely difficult to pick it out from the brilliant background. As for h, I have never been able to see it reversed, with either tangential or radial slit.

The first attempt to photograph a prominence was made on April 1. With a radial slit the C line of the second order rose to a height of about 1'30" above the limb, while F could not be seen bright. I nevertheless decided to use this line, as no plates were at hand for the red end, so it was brought into the center of the field, and the jaws of the second slit closed up until little more than the F line passed through. An ordinary Seed plate, sensitometer number 26 x, was put in the plate-holder, and moved slowly across the end of the telescope by cords connected with an independent clock. At the same time the speed of the regular driving clock of the horizontal telescope was so altered that the sun's image moved across the slit with a velocity equal (or nearly so) to that of the plate. On developing the plate, a portion of a nearly circular disc, corresponding to the photosphere, appeared to have had about the right exposure, but the region above the limb was undertimed, and showed no prominences.

Of course for each plate many variables had to be recorded; all of the following were noted in every case: Number of plate; date; hour; position of prominence; line and order of spectrum employed; length of exposure; focus of observing telescope; focus of sun's image on the slit; width of slit; speed of driving clock; speed of clock moving plate. In addition to the number just mentioned (which is placed in the lower right-hand corner), each plate is marked in the upper right-hand corner with a number corresponding with one on the box from which it was taken. A record of these box numbers is kept, giving: the size of plate; date received; name of maker; number of emulsion; sensitometer number; and remarks, stating dye (if any) used, and process of dyeing, etc. Thus by noticing the two numbers upon any negative, its complete history can be obtained.

With so many variables, each having its peculiar influence on the photograph, it is evident that the effect of any one can only be obtained by making the others constant in a series of photographs, and noting the changes produced by giving different values to the variable under consideration. The amount of light falling upon the plate depends upon the width of the slit, while the time of exposure of any point on the plate to this light is inversely proportional to the speed of the clock moving the plate. But this speed must be such that the plate and sun's image will move together, and is therefore proportional to the speed of the telescope's driving clock. Finally, the rate at which the sun's image travels across the slit determines the time during which light from a prominence passes through the slit, and falls upon some portion of the plate.

The best values of all these variables were obtained by making several series of photographs in the manner described, using in most cases the F line in the second order. For a given plate, a change of line means a change of focus of the telescopes, as well as a different exposure. There seems to be some latitude in the possible width of slit, altho of course a narrow slit is desirable, as the ratio of brilliancy of atmospheric to bright-line light is much decreased. But values of the width from one thousandth up to several thousandths of an inch can be used, thus changing the time of exposure without altering the clocks.

On April 14 a cool breeze was blowing, making the seeing fair in spite of a little whiteness in the sky. A hasty examination of the limb discovered a prominence in good position for the work, and a photograph was made through F, the slit being about .0015 inch wide. On developing the plate, the outlines of two prominences could be seen rising above the limb. As only one prominence had been noticed in observing the point in question, I returned to the telescope, and found that there were in fact two prominences in the exact positions shown in the photograph.

In this plate (A 45) the exposure was about right, and the focus excellent, so that the limb is sharply defined, tho it is of too great radius of curvature, owing to a slight difference in speed of the plate and sun's image. The region above the limb is somewhat fogged by the atmospheric light, but the two prominences can be clearly seen, tho of course only in their general form.

A word must be said in regard to the difficulty of obtaining a perfectly steady motion of the photographic plate. The friction of the guides prevents the carriage from starting until a certain force has been applied. When the pull becomes sufficiently great the carriage suddenly moves forward, then comes to rest again, and remains stationary until the same thing is repeated. At first a small independent clock was used to pull the carriage against a weight attached to its other end. But the clock was not strong enough for the purpose, and a weight was then made to do the work, the clock regulating the motion by unwinding a cord from a cone pulley. Photographs made with this arrangement were crossed with numerous vertical lines, due to inequalities in the clock's motion, and stretching of the cords. So the driving clock of the horizontal telescope was made to turn a small counter-shaft fixed to the table, and a cone pulley on this shaft conveyed its motion to the carriage through a fine wire, which thus replaced the cord first used. A piece of cat-gut carrying a weight was passed over a pulley suspended from the ceiling, and fastened to the upper end of the carriage. This gave the necessary tension, and secured a fairly constant motion, tho most of the plates are crossed by more or less lines. The photograph of the spectroscope shows the arrangement. [See page 142.]

As a greater uniformity of motion is more easily obtained with a higher speed, we have an argument for the use of as wide a slit as possible, allowing the plate and solar image to move more rapidly. In practice the width of slit is regulated by the density of the negative above the limb.

During the rest of April and the first two weeks of May there was a great deal of hazy weather, and even when perfectly clear there were no prominences of any brilliancy in favorable positions for work. A large number of exposures were made, however, but none of the plates showed more than the general form of the prominence.

Photographic Work with the Less Refrangible Rays

A number of experiments have been carried on with several organic dyes, in the hope of obtaining plates of sufficient sensitiveness to the red and yellow rays. Cyanine was first tried in the method employed by Burbank, and described by him as follows:

"Fifteen grams of cyanine are gently heated (over a steam bath) for from 30 to 40 minutes with 1 oz Chloral hydrate and 4 oz water. The whole mixture should now be stirred vigorously. While this operation is going on, 120 grams sulphate of quinine are dissolved by heat in a few ounces of a solution of 90% alcohol and 10% wood spirits. One ounce of strong aqua ammonia is now slowly added to the cyanine mixture above. Violent ebullition takes place immediately, chloroform being evolved, and cyanine is deposited in a soluble form on the sides of the vessel. After decanting off the supernatant liquid, 3 or 4 ounces of the above mixture of alcohol and wood spirits are added to dissolve the cyanine; the quinine solution is then added; and to the whole, more of the alcohol mixture until the whole measures from 8 to 9 ounces. This constitutes the "stock" solution, and should be kept away from all light, as it is very apt to become decomposed.

"All the above operations should be conducted in as little light as possible. The following straining and drying processes should be conducted in absolute darkness.

"To 30 oz water are added $1\frac{1}{2}$ drachms of the cyanine stock solution; the graduate that contained the cyanine is now washed out, $1\frac{1}{2}$ drachms of strong aqua ammonia are added, and the whole mixture is stirred vigorously. Into this bath two or three plates, or half a dozen strips, can be dipped at once. They should be left there about four minutes; meanwhile rocking the tray continuously so as to insure a uniform action of the dye."[16]

All of the operations were carefully carried out as directed, Seed plates of sensitometer number 26 being used, and developed in total darkness with a pyro developer of the following composition: —

Sodium sulphite	2 oz
Sodium carbonate	2 oz
Distilled water	32 oz
Alcohol	$1\frac{3}{8}$ oz
Salicylic Acid	$\frac{1}{4}$ oz

The salicylic acid is dissolved in the alcohol and added to the others.

One mustard spoon of dry pyrogallic acid with every 2 oz of the above makes an excellent developer, such as has been used in all my photographic work. An aqueous solution of bromide of potassium may be used as a restrainer, but my own experience tells me that proper dilution with water is more convenient, and just as efficacious, perhaps even more flexible.

Photographs made with the cyanine plates show them to be quite sensitive to the red rays, but hardly sufficiently so for my purpose.

The next dye tried was a soluble variety of alizarine blue. This was made up in a 1:10,000 aqueous solution, to which 1% strong aqua ammonia was added. This addition changed the brownish solution to a deep blue, which gradually becomes a light green if left exposed to the light. So the plates were stained in total darkness immediately after adding the ammonia. They were left in the bath about three minutes, and then dried in the dark.

Plates treated in this manner were somewhat more sensitive to the spectrum near C than were the cyanine plates, but still they would hardly do for prominence work.

These experiments with dyes having proved rather unsatisfactory, it was resolved to try the emulsion used by Abney for his photographic work in the infra red, as the curve given in his paper would indicate a considerable sensitiveness to the region including C.[17] The work of preparing the emulsion and coating the plates was undertaken by the foreman of the Harvard Dry Plate Co. After much delay the process was fairly started, when an explosion occurred while adding the solution of silver nitrate to the bromized collodion. The collodion caught fire, and all but the solids was consumed. As this took place on May 8, the experiments were reluctantly given up for the time.

The dye erythrosine was used for work on D_3 line. The use of erythrosine was introduced by myself at the Harvard Observatory in 1888, and it has been found very valuable in extending the limits of star spectra. The following formula gives the most satisfactory results: —

(a)
Erythrosine   50 m.g.
Distilled water   50 c.c.

(b)
Silver nitrate   50 m.g.
Distilled water   50 c.c.

Add (a) and (b) together, and then add 100 c.c. distilled water to the whole. Bathe plates in this 2 minutes, allow to dry an hour or so, and then bathe 1 minute in distilled water, finally leaving plates to dry in total darkness.

Plates thus treated are quite sensitive in the violet and blue, and so far as the D lines, tho they show a marked minimum in the green. They have enabled me to obtain very good photographs of D_3 by using a narrow tangential slit, but as no bright prominence was visible, they could not be tested for prominence photography.

Conclusion

Altho I have not been able in these limited experiments to produce photographs of any intrinsic value, nor to realize in any degree the possibilities of the photographic process, it is at the same time true that some results have been obtained. In spite of the insurmountable defects of the horizontal telescope, and in spite of the poor weather and limited number of prominences, the photographs demonstrate at least the feasibility of the method employed. Given a good refracting equatorial and a plate very sensitive to the longer waves of light, and I am confident that the spectroscope and attachments described in this paper will be sufficient to produce prominence photographs of real value for study and measurement.

This work I hope to continue at an early date. Meanwhile I shall pay special attention to the purely photographic questions involved, and to discover if possible some more efficient dye or sensitive salt. I am also anxious to try the method (b), as it seems in theory to have many advantages over (a), tho probably these would be somewhat reduced in practice.

The photography of the line D_3 tho amounting to little in itself, is perhaps new, and at least testifies to the value of erythrosine as a dye.

In closing, I wish to once more express my great obligations to the Director of the Harvard Observatory, as it is through his kindness that any experiments have been rendered possible.

Notes

1. J. Norman Lockyer, "Spectroscopic observations of the sun." *Proceedings of the Royal Society, XV,* 256-258, 1867.

*2. J. Norman Lockyer, "Notice of an observation of the spectrum of a solar prominence." *Proceedings of the Royal Society, XVII,* 91-92, 1869.

*3. Charles A. Young, "A New Form of Spectroscope." *Nature, 3,* 110-113, December 8, 1870.

*4. Friedrich Zöllner, Paper in *Report of the Royal Saxon Academy of Sciences,* February 1869. Translated by A. M. Mayer, "Observations of the Solar Protuberances." *Journal of the Franklin Institute, LVIII,* 317-319, 1869.

5. William Huggins, "Note on a method of viewing the solar prominences without an eclipse." *Proceedings of the Royal Society, XVII,* 302-303, 1869.

6. Angelo Secchi, *Le Soleil.* Vol. 1 (Paris: Gauthier-Villars, 1875), p. 232.

*7. J. Norman Lockyer and George M. Seabroke, "On a new method of viewing the chromosphere." *Proceedings of the Royal Society, XXI,* 105-107, 1873.

*8. Charles A. Young, "Observations of the solar protuberances." *Journal of the Franklin Institute, LX,* 3rd series, 232a-232b, 1870.

9. C. W. Zenger [letter], *Comptes Rendus des Séances de l'Académie des Sciences, 88,* 374, 1879.

10. Ostwald Lohse, "Über einen rotirenden Spectralapparat." *Zeitschrift für Instrumentenkunde, 1,* 22-25, 1881.

11. Ibid.

*12. Same as footnote 9, but is a quotation.

*13. Jules Janssen, "Sur la photographie de la chromosphère." *Comptes Rendus des Séances de l'Académie des Sciences, 91,* 12, 1880.

14. Heinrich Schellen, *Spectrum Analysis in its Application to Terrestrial Substances, and the Physical Constitution of the Heavenly Bodies.* Trans. from the 2d Ger. ed. by Jane and Caroline Lassell. Ed. with notes by William Huggins (London: Longmans, Green, 1872). See note by Capt. William de Wiveleslie Abney on p. 372.

15. L. Thollon, "Sur l'emploi de la lunette horizontale pour les observations de spectroscopie solaire." *Comptes Rendus des Séances de l'Académie des Sciences, 96,* 1200-1202, 1883.

*16. J. C. B. Burbank, "Photography of the least refrangible portion of the solar spectrum." *Proceedings of the American Academy of Arts and Sciences, 23,* 301-304, 1888.

17. Capt. William de Wiveleslie Abney, "On the photographic method of mapping the least refrangible end of the solar spectrum." *Philosophical Transactions of the Royal Society, 171,* 658-667, 1880.

*Quotations

Plate I.
The 12-inch reflecting telescope of
the Harvard College Observatory.

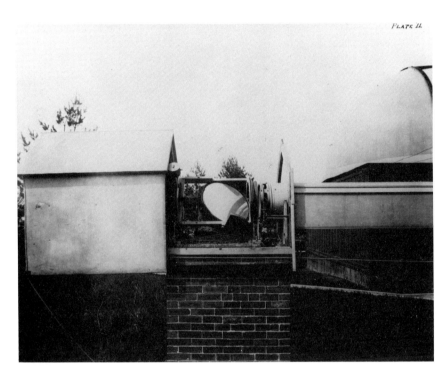

Plate II.
The 18-inch mirror.

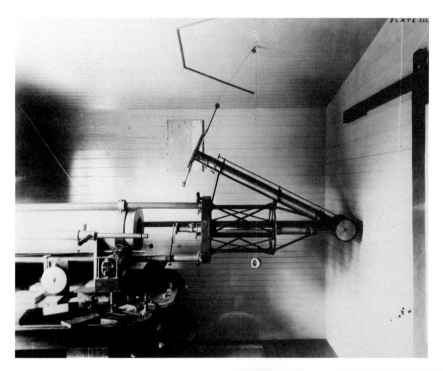

Plate III.
The spectroscope and the eye-end of the telescope.

Plate IV.
Sliding plate holder with adjustable slit.

The principle of the spectroheliograph. Slit (1) selects some particular segment of the solar image (see insert); slit (2) isolates a particular wavelength in the spectrum of that segment and allows it to impress its image on the photographic plate. As the sun's image is made to move across slit (1), the photographic plate moves in synchronism past slit (2). Thus a photographic image of the sun, in a particular wavelength of its radiation, is composed segment by segment. *(Courtesy of Sky and Telescope)*

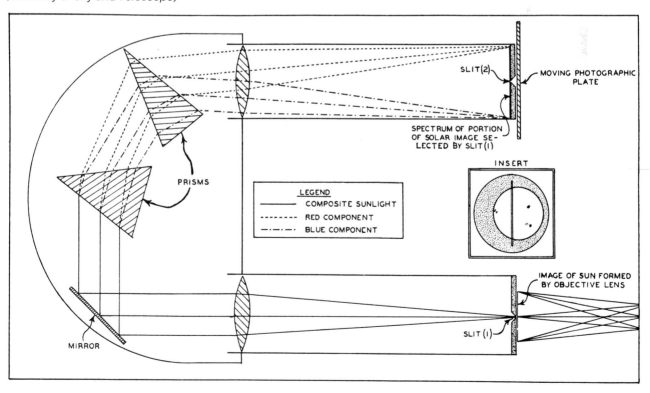

The new M.I.T. Campus, about 1916.
Massachusetts Institute of Technology.

A Plea for the Imaginative Element in Technical Education

There is strong reason for the belief held, with few exceptions, by our ablest university presidents that an institute of technology should be essentially a graduate school, in the same rank with schools of law and of medicine. For many years the best law schools have recruited their students from the graduates of colleges, and some of the leading medical schools have adopted the same principle. It has been felt that no amount of purely technical knowledge can replace the advantages of a broader culture and the better understanding of the affairs of the world which its possession implies. We need not pause to discuss here the relative educational value of science and the humanities, though this subject is touched upon in a later paragraph. Such weighing of one subject against another is not now relevant: we are concerned merely with the fact that students who have spent time enough to acquire a large amount of information of broad range are certain to have the advantage of those who have spent less time in acquiring less information of narrow range.

It is probable that the average member of a technological school is in more danger of a narrow outlook than any other class of students. In a large percentage of cases he has rejoiced from boyhood in a mechanical turn of mind, which has concentrated his attention on engines and machinery and the splendid achievements of modern engineering. Happy is the boy whose career is thus plainly foreshadowed. For him life is sure to be worth living, and the dangers of idleness may be ignored. But this very interest, in direct proportion to its intensity, is almost certain to lead to a neglect of other opportunities. The absorbing beauties of machine construction and design so completely occupy the boy's mind that they hinder a view of the greater world. He cannot be expected to perceive that a knowledge of the details of his chosen profession should not suffice to satisfy his ambition. He does not yet know that to become a great engineer he should cultivate not merely his acquaintance with the details of construction, but in no less degree his breadth of view and the highest powers of his imagination.

The greatest advances, whether in engineering, in pure science, in art, or in any other field, arise as mental pictures, at first uncertain as to details, but subsequently clear and distinct, requiring only an application of text-book methods to give them tangible form. It is in the conception of the picture, and not simply in the execution of the project it embodies, that the truly great engineer must excel. The mere dreamer never succeeds in bringing the confused and nebulous image to a sharp and definite focus. Lacking a substantial basis of knowledge, or otherwise failing to possess those subtle qualities which the realization of a splendid dream implies, he never gives walls or foundations to his castles in Spain. But practical ability to execute the design can never replace the design itself. The picture must be conceived and made visible to others before the work of construction can begin. Once the design has been transferred to paper and its fundamental principles made clear, an army of artisans, possessed of the skill required for its execution, can easily be found. It should be the purpose of the Institute to contribute to the world the largest possible proportion of men capable of conceiving great projects and the smallest possible proportion of men whose ambition can be completely satisfied by the work of executing them; and the means adopted to accomplish this end should be such as to improve the work of every graduate, including those who may be unfitted by nature for the greater tasks to which I have referred.

Perhaps it should be remarked at this point that what is ordinarily called invention, as applying particularly to machinery, is not alone considered here. A great engineer is not necessarily a great inventor, in this limited sense of the word. He may depend upon others to furnish the materials, whether perfected machinery or the simple brick or stone, copper or glass, with which he builds. It is for him to group them in such a way as to accomplish an advance, by securing greater economy in the industrial arts, by raising an architectural structure that shall benefit every occupant or casual observer, by facilitating transportation to such a degree as to revolutionize the conditions of daily life.

It would thus seem to be evident that a technological school can by no means afford to underestimate the need of broadening the view and cultivating the imagination of its students. What agencies, we may then ask, would best contribute to this end? It goes without saying that technical education must be the principal work of the school. Is it possible, in view of the heavy demands brought about by the rapid development of engineering, to give all necessary instruction in technical subjects, and also to extend the student's outlook upon the world and to develop his imaginative power?

I believe that three means contributing toward the accomplishment of this result should be considered: —

1. As a probable development of the future, the requirement of at least two years of general college work for entrance.

2. As a partial alternative under existing conditions, the allotment of as much time as can be spared to general studies in the Institute's curriculum, and the creation of new opportunities, outside of the regular work, for developing the social and cultural sides of the student.

3. As essential needs under all circumstances:

(*a*) Insistence upon the paramount importance of fundamental principles, as distinguished from specific facts and technical details.

(*b*) The fullest possible recognition and use of the educational value of science, both in its cultural aspects and in the means it affords of developing the reasoning powers and the constructive imagination.

Let us consider these points in the above order:—

1. It may be taken for granted that the progress of engineering will cause more and more difficulty in providing suitable technical instruction in a four years' course. Although I believe this difficulty can be partly met by giving less time to the mere acquirement of knowledge and more to practice in the solution of new problems, it is evidently no simple matter to reconstruct the curriculum on this basis.

The development of the turbine engine, for example, must be recognized in the course of instruction. Its adequate treatment, however, demands time, which can be had only by eliminating other instruction. So with the theory of alternating current machinery, the phenomena of radioactivity, and many other subjects of recent development. All must find place in the curriculum, which accordingly becomes more and more difficult and condensed. The increasing entrance requirements tend to shift the more elementary mathematical courses from the Institute to the preparatory school, and the same may be said of other subjects. The inevitable tendency is, therefore, for the purely technical courses to crowd out other work. At Sibley College this process has eliminated even modern languages from the curriculum. At the Institute political economy, English literature and composition, history, modern languages, and business law are retained, and successful efforts have been made to provide for much general reading through the adoption of requirements for summer work.

It may be expected, then, that the future will see the best of the technological schools requiring part, at least, of an ordinary college course for entrance. Such a result is earnestly to be desired, in view of the better and broader education rendered possible by such means. The technological schools may then devote themselves to professional studies, though pure science should always play a very important part in their work, and every effort should be made to realize and develop the more truly educational possibilities of the instruction. The rapid increase in the number of college graduates at the Institute, and the establishment of a three years' course for them, leading to an M.S. degree, are significant signs of the times.

2. We are told, however, that the average student is not in a position to spend six or eight years after leaving the preparatory school, in obtain-

ing an education. Without attempting to question the truth of this assertion, the analogous case of the medical schools seems to indicate that room might now be found for one or two technological schools requiring two years of college work for entrance. Nevertheless, I do not favor the immediate adoption of such a policy by the Institute. Further experience will show whether so radical a departure is essential. For the present we may consider the ordinary course limited to four years, and inquire whether it is possible to improve it in any considerable degree.

It may be hoped that the successful efforts made by the Faculty to retain a considerable number of general studies will be followed by an attempt to extend the scope of this work. The Institute graduate is in no less need than the Harvard graduate of a knowledge of history, literature, language, and art. The fact that the one may engage in engineering, while the other devotes himself to some other business, should draw no line of distinction between them. The engineer should know the accomplishments, the thoughts, and the ways of the world no less thoroughly than they are known by the broker, the banker, or the dealer in real estate. His work, as we have said, is not confined to the application of certain formulae to the solution of engineering problems. It occupies, or should occupy, a broader field, in which an understanding of the impelling motives and the probable actions, under given conditions, of other men is one of the first essentials of success.

The time will inevitably come when the mass of engineering knowledge which must be taught in some form in a four years' course will be double or treble what it is to-day. What can be done then? Will it not be possible, through the elimination of the less important details and greater concentration of attention on fundamental principles, to overcome the difficulty? If so, it seems reasonable to suppose that something of the sort could be accomplished now, leaving time for the inclusion of more general studies in the curriculum and for more practice in the solution of problems new to the student, by which his creative and reasoning faculties would be developed.

3. The saving of time should not be the only result of such reconstruction. There is reason to believe that the average student, at the present day, may often fail "to see the wood for the trees." His mind is not able to distinguish with sufficient clearness between fact and principle. A fact may be capable of attractive and forcible illustration, easily appealing to the mind. It may perhaps afford a most striking example of a general law, but the uninviting aspect of the latter, when reduced to a formula, may repel rather than attract. The law is soon forgotten, while the illustration of its application to a particular case is kept in mind.

But how, it may be asked, are we to escape the difficulty into which we have fallen? It is held, on the one hand, that double advantage may result from even greater attention than is now given to fundamental principles. It is admitted, on the other, that such principles must, in the nature of things, be taught and rendered attractive through just such illustrations as are now so effectively employed. Standing between the horns of this dilemma, we can only appeal for assistance to those who have demonstrated their ability in building up the Institute courses. In asking of them whether the last word has been said on this subject, we may confidently expect a negative reply, for the frequent revision to which the courses are subjected demonstrates a determination to keep abreast of the times. It may be hoped that this reference to the question will not be taken as a trivial attempt at criticism, since in their review of the year's work the members of the Faculty would probably have in mind the query here proposed.

It is undoubtedly true that no amount of general study and no method of teaching science can replace the advantages of personal experience. On the other hand, it must be admitted that, by adopting the best means to acquaint the student with the broader aspects of science, results may be accomplished which might otherwise be long delayed. The catalogue of the Institute rightly states, in opening its discussion of the courses of instruction, that the "fundamental elements in the curriculum of the school are mathematics, chemistry, and physics." It adds, further, "Instruction in technical methods is subordinated to the question of principles, and these principles are studied with the predominant purpose of exercising the powers and training the faculties." It would be difficult to prepare a more admirable statement of the purposes of the school, and this may seem to render any recommendations in this direction superfluous. As in the case of general studies, however, where these remarks may do no more than second the efforts already made by the Faculty, I may be permitted to emphasize the importance of extending the application of a principle already recognized and of adopting any practicable means of widening the outlook of the student.

In remarking upon the desirability of cultivating the scientific imagination and of developing that breadth of view which is most effectively acquired through reflection and experience, I have had in mind the fact that the most fertile and inspiring of all scientific theories has never, it would appear, received adequate recognition in the curriculum of educational institutions. I refer to the theory of evolution which, when rightly appreciated in its broadest scope, is perhaps better competent to awaken the imaginative powers and to develop an understanding of the greatest aims of science than any other single conception. Many institutions, the Institute among them, give a limited number of undergraduate courses involving the study of evolution in one or more of its innumerable phases. The opportunity remains, however, to present a general course of lectures dealing with evolution as applied to the various branches of science, and to require that it be attended by all students. Such a course, if accompanied by collateral reading and illustrated by a small museum of carefully selected objects, would do more, in my opinion, to accomplish the purpose in mind than any other single agency.

The natural tendency of the student, from which few escape, is to regard science as partitioned off into compartments, each more or less sufficient unto itself. Every effort should be made to break down this tendency, in order that it may become clear that science should be considered as a whole, particularly if its fullest educational value is to be realized. The theory of evolution, on account of its endless range and its importance in almost every branch of science, may serve as the best means of illustrating the arbitrary nature of the boundary lines that have been drawn. Even in the conception of evolution itself there is a natural but undesirable inclination to distinguish, for example, between organic and inorganic evolution, and to confine general courses of lectures to one or the other branch. What the student needs, if this view of the subject be correct, is some such picture of the general operation of the evolutional principle as Spencer has outlined, but so modified as to deal with the advances of recent years, and illustrated by the best and most striking examples that can be brought together.

Such a course of lectures should be arranged on a chronological basis, and would therefore open with a popular account of recent investigations on the origin and development of the heavenly bodies. The remarkable results of recent astronomical photography afford the richest of illustrative material for such lectures as these. Nothing could be more suggestive than the magnificent whirlpools of the great spiral nebulae, which are now considered as the source from which solar and stellar systems are developed. After seeing for himself the forms of these star sources, the student would listen with interest to an account of Laplace's nebular hypothesis and the more recent views which promise to supersede it. Then would follow a description of the sun as a typical star, and a sketch of stellar growth and development based upon modern inquiries. Here the intimate relationship between this field of astronomical research and the laboratory studies of the physicist and chemist would become apparent. For it is possible to solve physical and chemical problems through observations of the stars, as well as to solve solar and stellar problems through experiments in the laboratory. It would be easy, therefore, to introduce at this point such a sketch of modern physical and chemical conceptions as would bring home to the student the fundamental character of these branches of science, their relationship to other branches, and their remarkable development in recent years.

The transition to the next phase of the evolutional subject would be so natural as to be imperceptible. The formation and development of the earth and of its surface phenomena, which it is the function of the geophysicist and the geologist to study, involve another application of physical and chemical principles. At the present time the processes by which the rocks of the earth's crust were formed are being imitated in the laboratory, just as solar and stellar conditions are being reproduced, in minor degree, by laboratory experiments designed for the interpretation of astronomical observations. The picture of geological phenomena would be no less striking. What better mode of developing the scientific imagination could be found than that afforded by the conception of the early history of the earth? The rise and fall of mountains and continents; the changing area of the sea and the story of sedimentary deposits revealed in the stratified rocks; the growth of glaciers and the part they have played in former ages; the changes of climate; and, finally, the mysterious origin and development of plant and animal life, as first illustrated in the fossils, — such a picture as this, if viewed as a part of the greater picture which represents evolution as a whole, should stimulate the student to further inquiries.

Having advanced so far, he would eagerly await the account of the origin of species which can be given to such great advantage in the light of recent research. On the one hand there would be the evidence afforded by the countless specimens preserved in the rocks from former times, best typified perhaps in the case of the horse, whose many-toed ancestors can now be seen in an almost unbroken series, thanks to the energy and skill of recent investigators. On the other hand, even more attractive through the promise they hold out of future advances, the laboratory studies of experimental evolution, now pursued by both botanists and zoologists, would receive consideration. The splendid conceptions of Darwin and their brilliant exposition by Huxley; the clash of rival hypotheses which has since followed; the part played by natural selection, on the one hand, and by mutation, on the other, — these and many other aspects of evolution, from the botanical and zoological standpoint, are interesting in their popular appeal and of the highest value in developing breadth of view. In all of these lectures the personal side should not be forgotten. What better stimulus could be offered the student than that arising from an acquaintance with Darwin, in the quiet surroundings of his home, removed from the centres of intellectual activity, hampered by constant illness, and yet pursuing long and patiently those simple yet remarkable researches which formed the basis of "The Origin of Species"? And what a splendid contrast is afforded by the striking successes of Huxley, won in the midst of the turmoil of London, under the constant pressure of innumerable public duties! Here is an illustration of the most convincing kind that a scientific man is not of necessity a recluse, and of the more important fact that his work touches upon the concerns of the everyday world.

I might go on to develop, in greater detail and in clearer outline, my notion of the character which such a course of lectures should assume. Obviously, its limit need not be placed at the boundaries of organic evolution. It is much for the student to form a mental picture of the unity of science and of the orderly progress which culminates in the development of man. But, having pursued to this point the evolutional idea, he might wish to follow it further. The origin and growth of society, the course of history, the crude beginnings and the subsequent refinements of language, — in short, the source and progress of every form of material and intellectual activity are never to be rightly perceived unless in the clear light which the theory of evolution radiates.

I believe that this single example of the many agencies that might lead to the expansion of the student's intellectual horizon is one worthy of adoption. If science is to be regarded as not inferior to the humanities in its educational possibilities, it is because it deals with the largest and most fruitful conceptions, of which evolution is perhaps the greatest. While I am not of those who believe that science alone is competent to supply all of the student's needs or to take the place of the humanities in a well-rounded education, I would nevertheless maintain that, when rightly taught, it may do far more than merely to instruct the student as to the mechanism and the detailed mode of operation of the processes of nature.

It is unnecessary to remark on the uselessness or even danger of encouraging the growth of the imaginative power unless the power of reason and the capacity to carry projects into practical effect are developed in equal proportion. There is no occasion to fear that the practical side will suffer, for it is, and must always remain, the most conspicuous part of the Institute's work. Nor is there any chance that the imagination in such surroundings will outgrow reasonable bounds. It is nevertheless well to remember that no amount of imagination can replace a lack of common sense. Moreover, the necessity of discriminating between projects that are likely to work out well in practice and those that are merely ingenious, while devoid of genuine merit, must always be borne upon the student's attention. Sound training and severe practical experience must furnish the required criteria.

I have devoted so much attention to this plea for the needs of the undeveloped imaginative faculty that I may be suspected of underestimating the predominant importance of the Institute's technical work. Far from believing, however, that the school should deal with pure science to the detriment of applied science or with the humanities in such a manner as to stand in the way of the effective training of the engineer, I would support a movement which might extend still further the scope and the importance of the technical departments. The rapid development and brilliant success of the Research Laboratory of Physical Chemistry are well known. I believe not only in the establishment of such a laboratory in connection with the department of physics, but also in those departments which are more directly concerned with industrial progress. The recent suggestion of a laboratory of industrial chemistry, in which investigations bearing upon the needs so constantly encountered in the application of chemistry to the arts could be conducted, should receive careful consideration. The marked success already achieved by the Sanitary Research Laboratory and Sewage Experiment Station illustrates the possibilities of such a case. In engineering as well there is room for similar developments. It would seem that the plans already projected for graduate work in engineering should prove of great importance in the future development of the Institute.

It is pleasant to picture the possibilities that seem to lie so near at hand. Removed to another site, with a group of buildings expressive of the broadened scope of the new Institute and so attractive architecturally as to impress the student in his daily round, the school would be far better able than at present to compete with its powerful rivals. The provision, wherever feasible, of a separate building for each department, in which the fundamental purposes of the department might find full expression; the emphasis placed on the facilities afforded for the broadest possible education and the greatly increased opportunities for graduate and research work; the maintenance of the closest relations with manufacturing and industrial interests; the correlation, so far as possible, of all the researches carried on in the Institute, by instructors and students, with reference to their bearing upon large investigations of general importance; the formation of small but carefully stocked synoptic museums illustrating, on the one hand, the work of the various departments and, on the other, a course of lectures on evolution such as has been outlined; and, finally, the improvement of the student's daily life and associations by such agencies as will be afforded by the Walker Memorial Building, — by these means, and by others of like character, the Institute should be enabled to maintain its high reputation and to develop in such a manner as to satisfy the best ambitions of the alumni.

The suggestions offered in this paper may be summarized as follows: —
1. The need of greater breadth of view on the part of technical school graduates is likely to lead, in the best institutions, to the requirement of at least two years of general college work for entrance.
2. At present, on the basis of the existing entrance requirements and a four years' course, the policy of providing for more general studies

and of developing the student on the social and cultural sides should be continued and extended.

3. The efforts which are being made by various members of the Faculty to lay special stress on the importance of fundamental principles should receive hearty encouragement.

4. The fullest possible advantage should be taken of the educational value of science. A course of lectures on evolution is recommended as one of the most promising means of broadening the student's conception of science and of stimulating his imagination.

Solar Vortices and Magnetic Fields

I heartily appreciate the privilege of describing in this lectureroom some of the recent work of the Mount Wilson Solar Observatory. Like so much of the scientific research of the present day, it goes back for its origin to the fundamental investigations of English men of science. The spectroheliograph, which tells us of the existence of solar vortices, is a natural outcome of the application of the spectroscope in astronomy, where Englishmen were foremost among the pioneers. The detection of a magnetic field within these vortices followed directly from Zeeman's beautiful discovery of the influence of magnetism on radiation — a logical extension of the earlier work of Faraday — and from the classic investigations of Crookes and Thomson on the nature of electricity. In reviewing these great advances, investigators in other lands must again and again wonder at the exceptional ability of the English mind to make fundamental discoveries. When these discoveries have been made, it is a comparatively simple matter to utilise them in many departments of science. Americans cannot fail to rejoice that they may share in the traditions of a race which counts among its members the men who have given the Royal Institution its fame.

It is customary to distinguish sharply between the observational and experimental sciences, including astronomy in the former. In physics or chemistry the investigator has the immense advantage of being able to control the conditions under which his observations are made. The astronomer, on the other hand, must be content to observe the phenomena presented to him by the heavenly bodies and interpret them as best he may. I wish to emphasize the fact, however, that the distinction between these two methods of research is not so fundamental as it may at first sight appear. In 1860 a laboratory in which experiments were conducted for the interpretation of astronomical observations was established by Sir William Huggins on Upper Tulse Hill. The advantage of imitating celestial phenomena under laboratory conditions was thus appreciated half a century ago. I shall indicate later how important a part such a laboratory plays in the work of the Mount Wilson Solar Observatory. I shall also show that in other ways the astronomer may advantageously follow the physicist, particularly in the choice of observational methods, and in the design of instruments of research.

Sun-spots were discovered as soon as Galileo and his contemporaries directed their little telescopes to the sun. In fact, ancient Chinese records indicate that spots of exceptional size had been detected by the naked eye many centuries before. Long after their discovery, the most diverse views were held as to the nature of sun-spots. Sir William Her-

An address given at the weekly meeting of the Royal Institution of Great Britain, May 14, 1909. *Proceedings of the Royal Institution of Great Britain, 19,* 615-630, 1909.

schel mentioned the uncertainty which had existed prior to his time, remarking that the spots had been variously described as solid bodies revolving about the sun, very near its surface; the smoke of volcanoes; smoke floating on a liquid surface; clouds in the solar atmosphere; the summits of solar mountains, uncovered from time to time by the ebb and flow of a fiery liquid; etc. In Herschel's own view, the spots are to be considered as the opaque body of the sun, seen through openings in the luminous atmosphere which envelops it. Indeed, he considered that the sun should be regarded as the primary planet of our system, and even suggested the probability that it is inhabited. "Whatever fanciful poets might say, in making the sun the abode of blessed spirits, or angry moralists devise, in pointing it out as a fit place for the punishment of the wicked, it does not appear that they had any other foundation for their assertions than mere opinion and vague surmise; but now I think myself authorised, *upon astronomical principles,* to propose the sun as an inhabitable world, and am persuaded that the foregoing observations, with the conclusions I have drawn from them, are fully sufficient to answer every objection that may be made against it."*

Sir John Herschel did not abandon the idea of an opaque solar globe, but suggested that hurricanes or tornadoes might account for the piercing of the two strata of luminous matter which ordinarily conceal this globe. "Such processes cannot be unaccompanied by vorticose motions, which left to themselves, die away by degrees and dissipate — with this peculiarity, that their lower portions come to rest more speedily than their upper, by reason of the greater resistance below, as well as the remoteness from the point of action, which lies in a higher region, so that their centre (as seen in our waterspouts, which are nothing but small tornadoes) appears to retreat upwards. Now, this agrees perfectly with that which is observed during the obliteration of the solar spots, which appear as if filled in by the collapse of their sides, the penumbra closing in upon the spot, and disappearing after it."

We now know that sun-spots are brighter than the brightest arc light, and that their apparent darkness is merely the result of contrast with the intensely brilliant surface of the photosphere. We also know that the sun is a gaseous globe, attaining a temperature of about 6000° at its surface, and perhaps millions of degrees at its centre. If we examine a large-scale photograph of a sun-spot we see that it consists of a dark central region, called the umbra, and a surrounding area, decidedly less dark, called the penumbra. The structure of a spot, as this admirable photograph by Janssen shows, is granular, like that of the

---

*William Herschel, "On the Nature and Construction of the Sun and Fixed Stars," *Philosophical Transactions of the Royal Society of London, 85,* 20, 1795.

photosphere. In the penumbra these granulations seem to group themselves more or less radially, as though under the influence of some force directed toward or away from the umbra. Unfortunately, direct photographs of the sun have not yet attained such perfection as to show the most minute details of sun-spots. To appreciate these, we must have recourse to the exquisite drawings of Langley, the truthful quality of which is recognised by every astronomer who has observed sun-spots under favourable conditions. We shall see that the characteristic structure represented by these drawings is repeated, on a far greater scale, in the higher regions of the solar atmosphere disclosed on recent spectroheliograph plates.

Since the time of Sir John Herschel, many astronomers have proposed vortex theories of sun-spots. One of the first of these is the theory of Faye, who supposed the whirling motion to be the direct result of the peculiar law of the sun's rotation. This law was discovered by Carrington, who found from observations of spots near the equator that the sun completes a rotation in about 25 days, while the motion of spots at a latitude of 40° indicated the time of rotation to be nearly two days longer. Thus, as the rotation period increases toward the poles, the photosphere at the northern and southern boundaries of a sun-spot must move at different velocities (assuming the law of the sun's rotation to be the same as that of the spots). This difference in velocity would tend to set up whirling motions, clockwise in the southern hemisphere and counter-clockwise in the northern hemisphere. Sun-spots, in Faye's opinion, are the visible evidences of such whirls.

This theory has had many supporters, but it is now generally agreed that the difference in the rotational velocity of adjoining regions of the photosphere is not nearly sufficient to account for the observed phenomena. Secchi, one of the most assiduous observers of solar phenomena, was strongly opposed to Faye's theory. He pointed out that about 6 per cent of the spots he observed gave some evidence of cyclonic action, but in the vast majority of cases such forms as Faye's theory seemed to demand were lacking. We nevertheless owe to Secchi a most striking drawing of a sun-spot vortex.

When the spectroheliograph was first systematically applied to solar research in 1892, many rival theories of sun-spots occupied the field. Since the function of this instrument is to photograph the phenomena of the invisible solar atmosphere, it might be hoped that the results would throw much light on the nature of sun-spots. For many years, however, this hope was not realised. The first monochromatic images of the sun were made with the K line of calcium. If we compare such an image with a direct photograph of the sun, made in the ordinary way, we see that the sun-spots are surrounded and frequently covered by vast clouds of luminous calcium vapour. These attain elevations of several thousand miles above the sun's surface, but they must not be confused with the prominences, which ascend to much higher elevations. When observed at the sun's limb, the bright calcium flocculi, as these luminous clouds are called, are so low, in comparison with the prominences, that they can hardly be detected as elevations. Thus our knowledge of the calcium flocculi must be derived mainly from the study of spectroheliograph plates, which show them in projection on the disk. I must not omit to mention, however, that the calcium vapour rises to the highest parts of the prominences, and that this higher and cooler vapour frequently indicates its presence on spectroheliograph plates in the phenomena of dark flocculi. These are relatively inconspicuous, however, and need not be discussed here.*

It soon appeared that the average photograph of bright calcium flocculi could not be counted upon to indicate the existence of definite streams or currents in the solar atmosphere. In 1903 the hydrogen flocculi were photographed for the first time. By comparing these flocculi with the corresponding calcium flocculi we see that, in general, dark regions on the hydrogen image agree approximately in form with bright regions on the calcium image. This might appear to indicate that hydrogen is absent in the regions where calcium is most abundant. An investigation of the question, however, does not lead to this conclusion. Dark hydrogen flocculi seem to mark those regions on the sun's disk where hydrogen is present as an absorbing medium, which reduces the intensity of the light coming through it from below. In certain areas, where the temperature is higher or the condition of radiation otherwise different, the hydrogen flocculi are bright. In many cases eruptions are in progress at these points, but in others the difference in brightness is apparently not the direct result of eruptive action.

The hydrogen flocculi, thus photographed with the lines $H\beta$, $H\gamma$, or $H\delta$, differ in many respects from the calcium flocculi. Not only do they usually appear dark, where the calcium flocculi are bright: their forms exhibit striking peculiarities, which are absent or much less conspicuous in the case of calcium. The appearance of the calcium flocculi resembles that of floating cumulus clouds in our own atmosphere, whose capricious changes in form reveal the operation of no simple law. But the hydrogen flocculi, on the contrary, exhibit a definiteness of structure in striking contrast to this appearance. Some of the photographs strongly remind us of the distribution of iron filings in a magnetic field, and suggest that some unknown force is in operation.

Such was the condition of the subject when the red $H\alpha$ line of hydrogen was first applied to the photography of the flocculi, on Mount Wil-

*Eruptive prominences are also recorded on the disk as bright flocculi.

son, in March 1908. The calcium and hydrogen flocculi had been studied for several years, and much had been learned as to their nature and their motions. It had been found, for example, that the calcium flocculi observe the same law of rotation that governs the motions of sun-spots, while the hydrogen flocculi apparently follow a different law, in which the decrease in the angular rotational velocity from the equator toward the poles is much less marked. The latter result is in harmony with the investigations of Adams, whose accurate measures of the approach and recession of the hydrogen at the eastern and western limbs of the sun offer but little evidence of equatorial acceleration on the part of this gas. For this and other reasons it had been concluded that the hydrogen shown in such photographs reaches a higher level than the vapours of the bright (H_2) calcium flocculi. The region of the atmosphere previously explored with the spectroheliograph was nevertheless confined (except in the case of eruptions and dark calcium flocculi) to a comparatively low level, lying within a few thousand miles of the photosphere. What might be expected if a still higher region could be satisfactorily photographed in projection on the disk?

The red line of hydrogen offered the means of disclosing the phenomena of this higher atmosphere. As it may not immediately appear why different lines, caused by the radiation of the same gas, should not give precisely similar photographs, a brief reference to the aspect of a prominence in the red and blue hydrogen lines may be advantageous. Here are two photographs of the same prominence, seen in elevation at the sun's limb, one made with $H\alpha$, the other with $H\delta$. As the red line is very bright, even in the highest regions, the photograph taken with its aid shows the entire prominence. $H\delta$, on the other hand, is relatively weak at the higher levels, and consequently only the lower and brighter parts of the prominence are well recorded when this line is used. If, now, we suppose ourselves immediately above such a prominence, at a point where we observe it in projection against the disk, it is evident that the character of the hydrogen lines must depend upon their brightness at different levels. As we know that, speaking generally, absorption is proportional to radiation, the amount of light absorbed in the upper part of the prominence will be much greater for $H\alpha$ than for $H\delta$. Hence the average level represented by the absorption of $H\alpha$ will be higher than the average level represented by $H\delta$, since the higher gases play a more important part in the production of the former line. We may therefore expect that photographs of the sun's disk, taken with the light of $H\alpha$, will show the dark areas corresponding to absorption in the prominences much more clearly than photographs taken with $H\delta$. Moreover, since $H\alpha$ is stronger than $H\delta$ in the upper chromosphere, in regions where no prominences are present, the *average* level represented by this line will, in general, be higher than that represented by $H\delta$. A comparison of two photographs of the sun's disk, made with the lines in question, will suffice to make this clear. This enormous group of prominences, stretching for several hundred thousand miles across the sun, is much more clearly indicated by Hα than by Hδ. In general, the hydrogen flocculi are stronger and more distinct when photographed with Hα, and there are some regions which appear bright with Hα and dark with Hδ. This latter peculiarity probably has an important bearing upon the similar behaviour of hydrogen in certain stars and nebulae, but a discussion of this question cannot be undertaken here.

The first of the Hα photographs gave strong hopes of a substantial advance in our knowledge of the solar atmosphere. The sharpness and comparatively strong contrast of these flocculi, and the evidences of definite structure and clearly defined stream lines which they revealed were highly encouraging. The work was begun during the disturbed weather of the rainy season, when the definition of the solar image is never of the best. On April 30, 1908, the first photographs were secured under the fine atmospheric conditions which prevail in the dry season. This direct photograph (Fig. 1) shows a small and insignificant group of sun-spots, which would not seem, without other indications, to merit special attention. The next photograph (Fig. 2) shows that an enormous calcium flocculus occupied this region of the sun, but its form was in no wise remarkable, and afforded no evidence of the phenomena brought to light by the Hα photograph (Fig. 3). The structure recorded with the aid of the latter line recalls Langley's sun-spot drawing, and suggests the operation of some great force related to the sun-spot group. The same cyclonic structure had been less satisfactorily recorded on the previous day, but a comparison of the two photographs failed to indicate such changes as motion along the apparent stream lines might be supposed to produce.

The close of the rainy season now permitted an active study of the Hα flocculi to be undertaken. Many photographs were made daily, and the almost constant association of apparent cyclonic storms or vortices with sun-spots became evident. During several months of the year in California an unbroken succession of clear days can be counted upon, so that the changes of a given vortex can be followed without interruption. The cyclonic storms were found to be of two principal types: the first associated with groups of spots and represented in such photographs as those of April 30 and September 2; the second associated with single spots, and resembling a simple vortex, as illustrated in the photographs of September 9 and October 7, 1908 (Fig. 4). The appearance of these simple vortices is such as to indicate rotation in a clockwise direction in the southern hemisphere, and in a counter-clockwise direction in the northern hemisphere (assuming the direction of motion to be inward towards the spot). However, this cannot be taken as a general law, corresponding to the law of terrestrial cyclones. Indeed, many instances have been found of closely adjoining spots, in the same hemisphere, and frequently in the same spot-group, having magnetic fields of opposite polarity, produced by vortices rotating in opposite directions.

In some cases, at least, these vortices seem to exercise a powerful attraction on the surrounding gases, as a series of photographs taken on June 3, 1908 illustrates. A long dark hydrogen prominence, first photographed in elevation at the sun's limb on May 28, had advanced half-way across the solar disc. It lay at the outer boundary of a well-defined vortex, centered on a sun-spot. This spot had been gradually separating into two parts, and on June 3 the separation was complete. The first photograph of a series of nine was made on this day at 4 h. 58 m. Several successive photographs indicated no appreciable change, but one taken at 5 h. 7 m. showed that the prominence was developing an extension toward the spot. At 5 h. 14 m. this had assumed the appearance illustrated in the next photograph, and 8 m. later, when the last photograph of this series was taken, the extension had almost reached the spot. It will be seen that it divided into two parts, which indicates that each umbra was a centre of attraction. The average velocity of the motion toward the spot was over 100 km. per second. Later photographs, made on the following days, show a ring of bright hydrogen surrounding the spots, suggesting that the comparatively cool hydrogen carried down into the spots was reheated and returned to the surface, after escaping from the lower end of the vortex. We thus seem to be observing some of the phenomena of an actual vortex in the sun. But it must not be supposed that cases of this kind are common. In many instances the hydrogen flocculi do not appear to move toward or away from spots, but undergo changes of intensity, as though the physical condition of the gas were constantly changing. But before proceeding further with a discussion of these sun-spot vortices, let us turn to another phase of the subject, which will afford much new information indispensable for this purpose.

We are all familiar with the effect produced by passing an electric current through a wire helix. The lines of force of the resulting magnetic field are parallel to the axis of the helix, and its intensity is determined by the diameter of the helix, the number of turns of wire and the strength of the current. We also know, from Rowland's experiment, that the rapid revolution of an electrically charged body will produce a magnetic field. Thus if a sufficient number of electrically charged particles were set into rapid revolution by the solar vortices a magnetic field should result. What warrant have we for assuming the existence of charged particles in the sun, and how could such a field be detected?

Let me pass rapidly in review a series of phenomena with which you are all familiar. Sir William Crookes showed in this lectureroom as long ago as 1879, that the negative pole of a vacuum tube sends out a stream of particles capable of setting a light wind mill in rotation, and deviated from their straight path when under the influence of a magnetic field. He has kindly consented to show the same tube again to-night; you now see the effect upon the screen. The recent work of Sir Joseph Thomson and others has proved that these are negatively charged particles, called "corpuscles" or "electrons," and that their mass is about $\frac{1}{1700}$ of the mass of an atom of hydrogen. Moreover, Thomson has shown that at low pressures these corpuscles are given off from a hot wire or from the carbon filament of an incandescent lamp. He has also demonstrated that this property of emitting corpuscles at high temperature is common to carbon and to metals, whether in the solid or in the vaporous condition. Thus we have warrant for the belief that the sun, composed of just such elements as constitute the earth, must emit great numbers of these corpuscles. As Thomson has estimated that the rate of emission of a carbon filament at its highest point of incandescence may amount to a current equal to several amperes per sq. centimeter of surface, we can hardly be mistaken in assuming the existence of still more powerful currents in the sun. The emission of negatively charged particles implies the emission of positively charged particles, but in laboratory experiments, because of unequal rates of diffusion or other causes, charges of one sign are always found to be in excess. We thus have reason to believe that powerful magnetic fields may result from the revolution of these particles in the solar vortices.

In seeking a means of detecting such fields, let us first recall Faraday's discovery of the effect of magnetism on light, made at the Royal Institution in 1846. This discovery relates to the rotation of the plane of polarisation of light when passed through a plate of dense glass in a strong magnetic field. Although Faraday, in what was said to be his last experiment, endeavoured to detect the effect of magnetism on the lines of the spectrum, he failed because the apparatus then available was not sufficiently powerful. In 1896, Professor Zeeman examined with a large spectroscope the two yellow lines emitted by sodium vapour in a flame between the poles of a powerful magnet. Observing in the direction of the lines of force, he saw that the sodium lines widened when the magnet was excited. Subsequently with more powerful apparatus, he found that a single line, when observed under the above conditions, is split into two components by a magnetic field. The distance between the two components is a measure of the strength of the field. But the most characteristic quality of these double lines, which distinguishes them from double lines produced by any other known means, is the fact that the light of the two components is circularly polarised in opposite directions. If, then, we encounter a double line in the spectrum of any substance, and suspect it to be due to a magnetic field, we must apply the test for circular polarisation.

The simplest means of testing for circularly polarised light is to transform it into plane polarised light by passing it through a quarter-wave plate or a Fresnel rhomb. In the case of a Zeeman doublet, we would then have issuing from the rhomb the light of the two components, polarised in planes at right angles to one another. A Nicol prism, standing at a certain angle, will transmit one of these plane polarised beams and cut off the other. Turning the Nicol through 90° will cause the component previously cut off to be transmitted, and the other to be stopped.

Consider a sun-spot at the centre of the solar disk, and suppose it to be produced by a vortex, the axis of which lies on the line passing from the eye of the observer through the spot to the centre of the sun. Under these circumstances, if a strong magnetic field is produced by the vortex, the spectral lines due to vapours lying within this field should be widened or transformed into doublets. Moreover, the light of the components of these doublets should be circularly polarised in opposite directions. This would be true if the spot vapours were emitting bright lines, identical in character with those emitted by a radiating vapour between the poles of a magnet. The experiments of Zeeman, Cotton, König, and others, show, however, that dark lines, produced by the absorption of the spot vapours, should behave precisely in the same way as bright lines.

The spectrum of a sun-spot was observed for the first time by Lockyer in 1866. He found that many of the lines of the solar spectrum were widened where they crossed the spot, and the observation of these widened lines has been carried on systematically by many observers ever since. Conspicuous among these observers was Young, whose last observations were made with a powerful grating spectroscope attached to the 23-inch Princeton refractor. This instrument showed that some of the spot lines are close doublets. Dr. Walter M. Mitchell, who at first worked in conjunction with Professor Young and later by himself, gave special attention to these double lines, which he found to be particularly numerous at the red end of the spectrum. He called them "reversals," and the existing evidence favoured the view that they were produced by the radiation of a hotter layer of vapours overlying the spot, which would give rise to a narrow bright line at the centre of the widened dark line. True reversals of this kind actually seem to occur in the case of H and K and other lines in the spot spectrum, and it was therefore natural that Mitchell should attribute the similar phenomena of the spot doublets to a similar cause. It was generally supposed that the widening of the dark lines was due to the increased density of the spot vapours. The diverse character of the lines in the sun-spot spectrum is well illustrated by this drawing, which is due to Mitchell.* In addition to the ordinary widened and "reversed" lines, we find cases where a dark central line is accompanied by wings, others in which lines are thinned or completely obliterated, etc.

I have already referred to the importance of applying in astronomical research the methods of the physicist. During the last quarter of a century the study of spectroscopic phenomena in the laboratory has been completely transformed. It may well be said that this transformation, which has involved such discoveries as spectral series, the effect of pressure on wave-length, and the Zeeman effect, has been directly due to the use of Rowland's concave gratings, of great focal length, arranged for photography. In astronomical spectroscopy great advances have also been made, but the spectroscope has continued to occupy the place it formerly held as an attachment of the telescope. Although Rowland used a long-focus concave grating for his classic study of the solar spectrum, the heliostat and lens employed with this instrument gave so small a solar image on the slit that the investigation of sun-spots and other details was impossible. We thus see that while in the observatory the spectroscope continued to be used as an accessory of the telescope, in the laboratory the parts were exchanged, and the telescope was employed simply as an accessory of the spectroscope. It seemed obvious that a great opportunity for advance lay open to the investigator who would combine a long focus spectroscope with a long focus telescope. As it would be difficult or impossible to use for photography a sufficiently long spectroscope attached to the tube of an equatorially mounted telescope, some form of fixed telescope was plainly essential.

The tower telescope on Mount Wilson (Fig. 5) is designed to accomplish this purpose. It consists essentially of a 12-inch refracting telescope, of 60 feet focal length, mounted in a fixed position, pointed directly at the zenith. The ordinary telescope tube is replaced in this case by a light steel tower, firmly held in position by steel guy ropes. The 12-inch objective lies horizontally at the summit of the tower, and sunlight is reflected into it from the second of two adjustable plane mirrors. The first of these mirrors is mounted as a coelostat, and is rotated by an accurate driving-clock about a polar axis at such a rate as to counteract the apparent motion of the sun. Thus a beam of sunlight is reflected from the coelostat mirror to the second mirror, which sends it vertically downward through the objective. In the focal plane, 60 feet below the objective, an image of the sun, about 6.6 inches in diameter, is formed on the slit of a spectrograph, at a height of about three feet above the surface of the ground. After passing through the slit, the light of any desired portion of the solar image (a sun-spot, for

---

*The drawing was apparently shown in Hale's lecture but was not included in the published article. [Editors]

example) descends vertically into a well about 30 feet deep, excavated in the earth beneath the tower. Thirty feet from the slit the diverging rays encounter a 6-inch objective, through which they pass. After being rendered parallel by the objective, the rays fall upon a Rowland plane grating, ruled with 14,438 lines to the inch. The grating breaks up the light into a series of spectra, and the rays are returned through the same objective, which brings the spectra to a focus at a point near the slit. By inclining the grating at a slight angle, the image of the spectrum is made to fall at a point slightly to one side of the slit, and here the photographic plate is placed. Thus a portion of the spectrum 17 inches in length can be photographed in a single operation. In the work on sun-spots, most of the photographs are taken in the third order of the grating, where the dispersion and resolving power are very high. When the spot spectrum is being photographed, only the light from the umbra is admitted to the slit. At the end of the exposure this portion of the slit is covered, and light from the photosphere, at a point removed from the spot, is admitted to the slit on either side. Thus the narrow spot spectrum is photographed between two strips of solar spectrum, used for comparison.

The advantages of this combined form of telescope and spectrograph are considerable. On account of the great thickness (12 inches) of the mirrors, the height of the coelostat above the heated earth, and the use of a vertical beam, the definition of the solar image is always better than with the Snow (horizontal) telescope. Another important advantage is the nearly constant temperature at the bottom of the well, where the grating is placed. This permits long exposures to be given, when necessary, without danger of such displacements of the spectral lines as would be caused by expansion or contraction of the grating. The grating used in this spectrograph is a small one, which I have employed in most of my work since 1889, but the unusual focal length of the spectrograph permits the full visual resolution of the grating to be utilised in photographic observations. Thus it has become possible to photograph the widened lines and doublets, as well as a host of narrow lines, most of them due to chemical compounds, which had not previously been recorded in the spot spectrum.

Lack of time prevents me from discussing in this lecture the various studies of sun-spot lines carried out with this instrument before the attempt to detect a magnetic field in spots was undertaken. An extensive catalogue of these lines is nearly complete, a preliminary map has been issued and a better one is in preparation, and a series of investigations with the arc and electric furnace has suggested that the strengthening and weakening of certain lines is due to a reduction in the temperature of the spot vapours. At present we are concerned with the cause of the widening and doubling of spot lines, and the method of testing this question must now be described.

A Nicol prism was mounted above the slit of the spectrograph, and just above this a Fresnel rhomb. If the components of a spot doublet were circularly polarised in opposite directions, passage through the rhomb should give two plane polarised beams, the planes of polarisation making an angle of 90° with each other. Thus in one position of the Nicol one of the components should be photographed alone, and by turning the Nicol 90° this should disappear and the other component come into view.

When this test was applied with the tower telescope, in June, 1908, the true character of the spot doublets became apparent. One or the other component of the doublet could be cut off at will by rotating the Nicol, precisely as Zeeman had done in the laboratory. On account of the unique character of the Zeeman doublets, this test alone was almost sufficient to prove the existence of a magnetic field in sun-spots. But one of the great beauties of the Zeeman effect is its many-sided character, which permitted the test to be multiplied and extended. From Zeeman's first experiments it was known, for example, that if the strength of the magnetic field is insufficient to separate completely the components of a doublet, the edges of the resulting widened line should be circularly polarised in opposite directions. Thus those lines which are widened, but not doubled, in spots might be expected to shift in position when the Nicol is rotated. This was found to be the case. Again, the lines which constitute the flutings of the spectra of compounds are not, in general, affected by a magnetic field. Hence such lines in the spectrum of a sun-spot should not be shifted when the Nicol is rotated. This, also, was found to be true. But a still more satisfactory test was suggested by another laboratory phenomenon. When a doublet is observed along the lines of force, with one of the components extinguished by the Nicol, reversal of the current through the magnet should extinguish the visible component and cause the invisible one to appear. In the sun, according to our hypothesis, reversal of the direction of revolution in a vortex should correspond to reversal of the current through the coils of a magnet. Hence the red component of a doublet should appear in the spectrum of a vortex rotating in one direction, the violet component in that of a vortex rotating in the reverse direction. Fortunately, the appearance, on opposite sides of the solar equator, of two spot vortices rotating in opposite directions (Fig. 4), made this test possible. The results were perfectly in accord with the hypothesis.

So far we have been considering only such phenomena as are observed parallel to the lines of force of a magnetic field. But a spectral line which, under such circumstances, appears as a doublet, is usually transformed into a triplet, when the observation is made at right angles to the lines of force. The circularly polarised side components of the doublet give place to plane polarised components, occupying the same position, while another line appears centrally between them. The light of this line is also plane polarised, the direction of the vibrations being parallel to the field, while the vibrations of the side components are in a plane at right angles to the field. Thus when a spot is carried by the solar rotation to a point near the limb, we might expect the double lines in its spectrum to be transformed into triplets, if produced by a magnetic field. The failure of the central line to appear seemed to raise an important argument against the magnetic hypothesis.

At this point the necessity of conducting laboratory investigations in immediate conjunction with astronomical observations is well illustrated. Fortunately, our laboratory was already well equipped for work of this nature (Fig. 6). In anticipation of the possibility that observations of the Zeeman effect would be needed in the interpretation of solar and stellar phenomena, a powerful electro-magnet, with suitable accessory apparatus, had been provided. A brilliant spark produced between metallic electrodes in the field of the magnet furnished the source of light. As many of the double lines in sun-spot spectra are due to iron, this metal was selected for the first experiments. The spectrum was photographed, at various angles with the lines of force, with a powerful spectrograph, like the one used with the tower telescope, similarly mounted in an underground chamber.

The difficulty of accounting for the behaviour of the iron doublet in the sun was removed by these investigations. It appears that these lines do not become triplets, when observed across the lines of force. In reality they are changed to quadruplets, or doublets in which each of the components is a close double line. In the magnetic field of sun-spots, which is much weaker than the field used in the laboratory, the closely adjoining lines which constitute the components of the doublets cannot be separated. Thus these sun-spot lines should appear double at whatever position the spot may occupy on the sun's surface.

The distance between the components of doublets or triplets separated in the magnetic field varies greatly for different lines. Some exceptional lines are not affected in the least, others are merely widened, and others are clearly and sometimes greatly separated. It is therefore important to compare the widening and the separation of lines in a sun-spot spectrum with the corresponding phenomena in the magnetic field. With few exceptions, most of which may be accounted for by the presence in the spot spectrum of closely adjoining lines of other elements, the solar and laboratory results were found to be in good agreement. The following table gives a comparison of certain iron lines in the spot and laboratory.

Wave-Length	Δλ, Spark	$\frac{\Deltaλ, \text{Spark}}{5 \cdot 1}$	Δλ Spot	
6213·14	0·703	0·138	0·136	−0·002
6301·72	0·737	0·144	0·138	−0·006
6302·71	1·230	0·241	0·252	+0·011
6337·05	0·895	0·175	0·172	−0·003

The column headed "Δλ, spark" gives the distance between the components of the lines as observed in the laboratory. As the strength of the magnetic field used in the laboratory was about 5·1 times that of the spot, the quantities obtained by dividing the separations in the second column by 5·1 are given in the third column. These separations are directly comparable with the separations of the corresponding lines in the spot, which are given in the fourth column. The fifth column shows that the differences between the solar and laboratory results are very small. As the strength of the field in the laboratory was about 15,000 gausses, the strength of the field in this spot would be about 15,000 ÷ 5·1 = 2900 gausses. The strongest field hitherto measured on our photographs of spot spectra is about 4500 gausses, corresponding to a considerably greater separation of the lines (Fig. 7).

When a similar comparison was made for various lines of titanium and chromium, a much less perfect agreement between the spot and laboratory results was found. It had already been observed that such lines as D of sodium and b of magnesium, which undoubtedly represent a much higher level than the great majority of lines in the spot spectrum, are but very slightly widened. As these lines are strongly affected by a magnetic field in the laboratory, it appeared evident that the strength of the field in spots must fall off rapidly in passing outward through the spot vapours. Under these circumstances lines of other elements, which represent levels higher than the average, should show small separations in the magnetic field of the spot. It seems probable that in this way the lack of perfect agreement between the laboratory and solar results, observed in the case of titanium and chromium, can be accounted for.

A further important test was afforded by the well-known phenomenon exemplified in Preston's law. According to this law, the distance between the components of the lines split up by a magnetic field varies directly as the square of the wave-length. This we found to be true even in the case of a metal like iron, the lines of which cannot be grouped into series, if the average separations of a sufficient number of lines were considered. We should therefore expect that the widening of lines in spots would rapidly decrease toward the violet and that the separation of spot doublets should diminish in a similar way. A study of the spot spectrum shows that this actually occurs.

It soon appeared that the normal spot spectrum always contains triplets as well as doublets (Fig. 7). These are less easily recognised, because the presence of the central line crowds the components so closely together that they are not readily separated with the resolving power available. As these triplets are photographed even when the spot is very near the middle of the sun, it is evident that the spot always sends out light which makes a considerable angle with the lines of force. In a normal triplet the central line is of twice the intensity of the side components, when observed at right angles to the lines of force, and disappears altogether when observed parallel to the lines of force. Thus, by determining the relative intensities of the central and side lines of such a triplet, the angle between the lines of force and the line of vision can be obtained. In the case of sun-spots, the data at present available are not sufficient for the accurate determination of this angle, but it seems to lie between 30 and 60 degrees, when the spot is near the centre of the sun. On the hypothesis that the magnetic field is produced by the spot vortex, it would then follow that the axis of the vortex, instead of being radial, as we at first assumed, makes an angle of much less than 90° with the surface of the photosphere.

The time at my disposal permits me to describe briefly only a few other phases of this investigation. In the laboratory the central line of triplets is polarised in a plane parallel to the magnetic field. Hence, if the light is passed through a Nicol prism, used without a rhomb, it should be possible to extinguish this line at certain positions of the Nicol, in which case a spot triplet would appear as a doublet. This test has also been applied to the spot triplets, with the expected result. In fact this method supplies a convenient means of recognising close triplets, the components of which are too closely crowded to be seen separately before the central line is cut out. Indications have also been obtained of what may prove to be unequal rotation of the plane of polarisation of this central line in different parts of spots. The gradual decrease in the strength of the field from the umbra to the outer limit of the penumbra has been studied, and magnetic fields have been detected on the sun's disk in certain regions outside of sun-spots. It is evident that many new phases of the subject are likely to be developed in the future, especially if larger images of the sun and more powerful spectrographs are employed. In this connection it may be stated that a tower telescope of 150 feet focal length, to be used on Mount Wilson, with a spectrograph of 75 feet focal length, is now under construction. This will give a focal image of the sun about 16 inches in diameter, in which small spots, as well as large ones, can be studied.

Although it now seems to be demonstrated that sun-spots are electric vortices, judgment should be reserved as to the various theories which have been advanced to account for their origin. Many of the results I have described appear favourable to Emden's solar theory, but it seems to be opposed by the important investigations of Evershed, who has found that the metallic vapours in sun-spots flow radially outward from the umbra, parallel to the photosphere. The further development of Evershed's work, and the continued study of solar vortices and magnetic fields, should soon permit a reliable theory of sun-spots to be formulated.

It is evident that the rapid decrease upward of the strength of the field in spots would prevent it from having an appreciable influence on the higher solar atmosphere. At the distance of the earth, as Schuster has shown, the combined magnetic effect of several spots, all assumed to be of the same polarity, and not taking into account such rapid decrease in strength at higher levels as is actually observed, would be altogether incompetent to account for terrestrial magnetic storms.

In concluding, I wish to express my appreciation of the assistance I have received from my colleagues at Mount Wilson. I am particularly indebted to Messrs. Adams, Ellerman, King, Nichols and St. John for aid in connection with the present investigation.

Fig. 1.
Direct photograph of sun-spot group 1908, April 30, 6 h. 25 m. A.M. Pacific Standard Time.
The Hale Observatories.

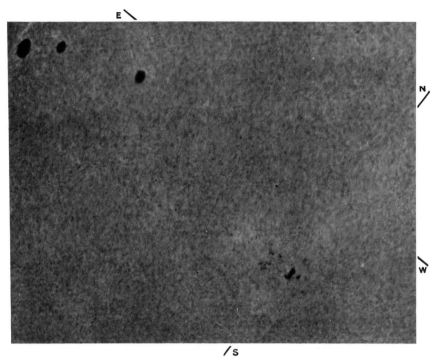

Fig. 2.
Same region of the sun showing the calcium (H$_2$) flocculi 1908, April 30, 4 h. 43 m. P.S.T.
The Hale Observatories.

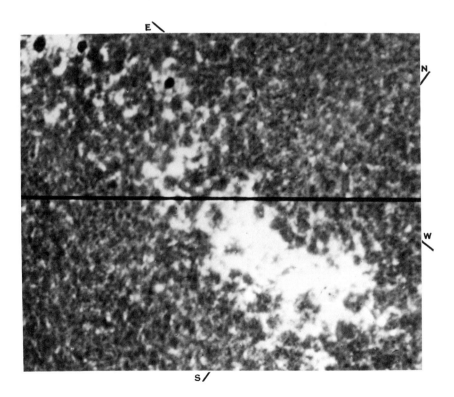

Fig. 3.
Same region of the sun, showing the hydrogen (Hα) flocculi 1908, April 30, 5 h. 06 m. P.M. P.S.T.
The Hale Observatories.

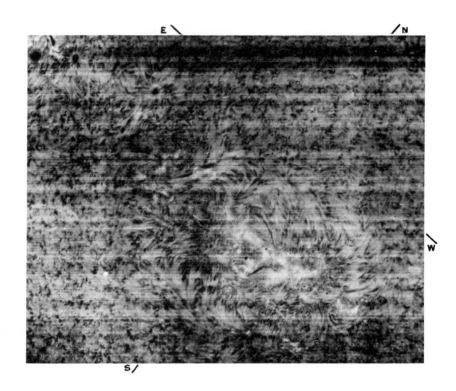

Fig. 4.
Sun-spots and hydrogen floculli showing right- and left-handed vortices 1908, October 7, 7 h. 02 m. A.M. P.S.T.
The Hale Observatories.

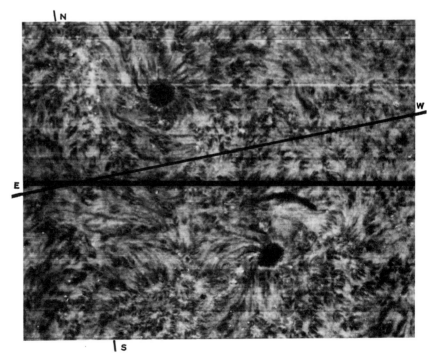

Fig. 5.
Sixty-foot tower telescope on Mount Wilson.
The Hale Observatories.

Fig. 6.
Interior of early physical laboratory on Mount Wilson, showing spectrograph and Magnet used in study of Zeeman effect.
The Hale Observatories.

Fig. 7.
Iron doublet (λ 6301·72) and triplet (λ 6302·71) in two spot spectra, showing field strengths of 2900 and 4500 gauss, respectively.
The Hale Observatories.

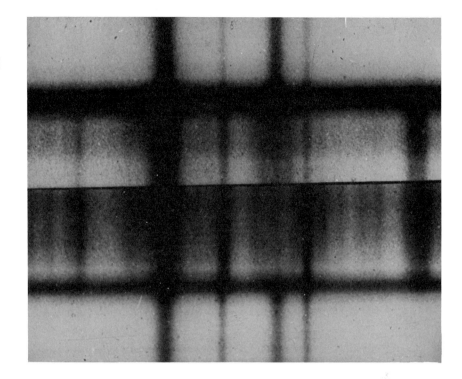

Meeting of the National Academy of Sciences in the 1870s.
The Smithsonian Institution Archives.

1. Joseph Henry
2. Mary Henry
3. William Jones Rhees
4. Frank Wigglesworth Clarke
5. John Strong Newberry
6. John Call Dalton
7. Julius Erasmus Hilgard
8. Joseph J. Woodward
9. Peter Parker
10. Alfred Marshall Mayer
11. William Ferrel
12. Benjamin Silliman
13. Clarence Edward Dutton
14. Emil Bessels
15. Arnold Guyot
16. John Huntington Crane Coffin
17. Benjamin Apthorp Gould
18. Elias Loomis
19. Charles Anthony Schott
20. George Engelmann
21. Benjamin Peirce
22. Simon Newcomb
23. Lewis Henry Morgan
24. Albert Abraham Michelson
25. John Shaw Billings
26. Silas Weir Mitchell
27. Endlich

National Academies and the Progress of Research

The Future of the National Academy of Sciences[1]

In previous papers of this series[2] we have traced the development of European academies and observed the powerful influence they have exercised on the advancement of research; we have watched the beginnings of scientific investigation in the United States, and their public recognition by act of Congress establishing the National Academy of Sciences; and we have followed the history of the Academy during the half century which has elapsed since its origin. In view of the great part which academies have played in the past, and the fact that the rapid development of original research in this country has carried us out of the pioneer period, the National Academy now faces an exceptional opportunity to impress its influence upon the future scientific work of the United States. But if it enjoys an opportunity, it also faces a duty, imposed upon it by its national charter and by its position as the sole representative of America in the International Association of Academies. The history of the Academy shows that it has taken its obligations seriously, by complying with requests from the executive and legislative departments of the government for advice on scientific matters, by the use of trust funds for the advancement of research, by the award of prizes and grants for investigation, by the initiation and support of international cooperation in research, and by such other means as its limited endowment has permitted. But while the rapid growth of the scientific bureaus of the government has reduced the number of questions which would otherwise be submitted to the Academy, the enormous increase in the wealth of the country, and the expansion of its trade relations have raised new problems and advanced new opportunities. These developments, which have resulted in the multiplication of universities, observatories and laboratories, and the foundation of great endowments for research, place the Academy in a new position, and impose the question whether it can not now accomplish much more than was formerly possible. It is the purpose of this paper to open the discussion of this question, in the hope that its further consideration by other members may lead to an extension of the work and usefulness of the Academy.

Fortunately we may take advantage of the rich store of experience accumulated by the European academies during their long histories. In seeking to adapt this to our own needs, we must of course recognize the special conditions existing in the United States. The great area over which our members are distributed and the lack of any such centralization as we see in London or in Paris, will always stand in the

Science, 40, 907-919, December 1914.

[1]This paper was presented at the Baltimore meeting of the National Academy in November, 1913. By action of the council, a manuscript copy was subsequently sent by the home secretary to each member of the academy for criticism and comment. In preparing the paper for publication, the author has had the advantage of seeing these replies. Except for a few minor verbal changes, the text is printed in its original form, with the addition of new paragraphs in square brackets.

[2] I. "The Work of European Academies," *Science, 38,* 681, 1913. II. "The First Half Century of the National Academy of Sciences," *Science, 39,* 189, 1914.

way of weekly meetings like those of the Royal Society and the Paris Academy. But if we can not hope to see our leading investigators personally demonstrate each step in their progress before academic audiences, as Faraday and Pasteur and many another have done abroad, we can nevertheless provide for lectures and papers illustrated by experiments in connection with the semi-annual meetings of the Academy, and possibly for others of a public character, extending throughout the year, after the manner of the Royal Institution of London. The disadvantage of our members in being unable to read accounts of their latest advances before weekly meetings of their colleagues can also be largely offset by the publication of *Proceedings,* in which the first results of all new work may be adequately presented. Thus, though we lack some of the advantages of centralization, these may be largely overcome, while retaining the very great advantage of a widely distributed membership representing the scientific interests of every section of the country.

Functions of a National Academy

The criticism has sometimes been directed against academies covering the whole range of knowledge that their place has been sufficiently filled by the special societies devoted to particular branches of science. For more than a century the Royal Society and the Paris Academy served all the purposes of science in Great Britain and France, but toward the end of the eighteenth century special societies began to develop in England. The establishment of the Linnean Society in 1788 did not appear to give special concern to the members of the Royal Society. But when the Geological Society was instituted in 1807, Sir Joseph Banks, then President of the Royal Society, united with Sir Humphry Davy and others in a strenuous attempt to amalgamate it with the parent body. The Royal Astronomical Society was established in 1820, partly as the result of the accumulation of valuable observations too extensive for the Royal Society to publish. Sir Joseph, though he had himself aided in the establishment of the Linnean Society, was greatly perturbed at this further development. A short time later he died in the belief that the special societies had struck a severe blow at the respectability and usefulness of the Royal Society, by robbing it of many of its members and laying claim to some of its most important departments.[3] But his fears were wholly unwarranted, and the special societies continued to grow and multiply, to the advantage of science and of the Royal Society itself. Their extensive publications have not detracted from the volume or the quality of the *Philosophical Transactions* and the *Proceedings,* and each of these societies, by contributing to the development of some special field, has helped to build up that great organization of British science of which the Royal Society is the acknowledged and venerated head.

[3]Sir James Barrow, *Sketches of the Royal Society and the Royal Society Club* (London, 1849), pp. 10, 256; C. R. Weld, *A History of the Royal Society* (London, 1848), pp. 242, 246.

These details will not be out of place if they help to emphasize a principle which should always be respected in the work of the National Academy. The societies and journals which have been established to meet the needs of scientific progress have come to stay. It is neither necessary nor in any way desirable to usurp their functions, which are the result of a natural process of evolution. There is ample room, however, for academies devoted to the whole range of science. The rapid advance of research in a thousand ramifying fields has left much intermediate territory unexplored. The approach to these undeveloped regions may be made from more than one direction, and through the aid of more than one method. Thus nothing can be more stimulating to the progress of research than an acquaintance with the investigations and processes which are constantly being developed in fields other than one's own. Mathematics has received its principal impulses from astronomy and physics. Physical chemistry is indebted, on the one hand, to Pfeffer the botanist for the study of vegetable cells, and on the other to the mathematical and physical investigations of Willard Gibbs, Van der Waals and Arrhenius. Astrophysics came into existence through the use in astronomy of the spectroscope and other physical instruments. Every department of science sheds a luster which should illuminate, not only its particular territories, but others, near and far, occupied by other workers. The importance of recognizing and utilizing this fact must therefore increase as time goes on.

[It has been truly said that an academy can hope to accomplish large results only as it succeeds in meeting the conditions of the present rather than those of the past. What are existing conditions in science? Surely none is more striking than the contraction of the field of the average investigator. Specialization is inevitable in the maze of modern progress, and the narrowing effect of constant devotion to a single subject must become still more apparent as science ramifies further. A general academy, by insisting on the importance of large relationships, by demonstrating the unity of knowledge, by recognizing the fact that fundamental methods of research, wherever developed, are likely to be applicable in more than one department, can do much to broaden and to stimulate its members. The correlation of research should be counted as one of its prime objects, and its energies should be largely directed to this important end.]

We are thus led to the conclusion that the functions of a National Academy should be of the broadest character, and that the advantage of sharing in the results of all its departments should belong to every member. Thus the policy of our National Academy of avoiding division into separate sections,[4] and of bringing papers on the most diverse

[4] Except for voting purposes.

subjects before the entire body, is fundamentally sound and should be maintained. Later in this paper the question will be considered whether the range of the Academy's activities should be extended so as to give increased recognition to departments of knowledge other than the physical and natural sciences.

Under the conditions now existing in the United States, there is reason to believe that the functions of a National Academy might well be multiplied so as to meet a wide variety of needs. It should stand, first of all, as a leading source and supporter of original research and as the national representative of the great body of American investigators in science. To the government it should make itself necessary by the high standard of its work, the broad range of its endeavors, and the sane and scientific spirit underlying all of its actions. To its members it should offer stimulus and encouragement in their investigations; due recognition of their advances; financial assistance and the use of instruments at critical periods in their work; the advantage of listening to papers ranging over the whole field of science, bearing suggestions of principles or methods likely to develop new ideas; contact with the greatest leaders of research from all countries and opportunities to listen to descriptions of their work; access to books and manuscripts not easily obtainable from other sources; and participation in international cooperative projects in every field of investigation. In the public mind it should rank as the national exponent of science, and as the agency best qualified to bring forward and illustrate the latest advances of its own members and of the scientific world at large. To representatives of manufactures and industries, the Academy should serve to promote the appreciation and widespread use of the scientific principles and methods which have built up the great industrial prosperity of Germany. With other societies devoted to various branches of science, it should cooperate in harmony with the best interests of American research. Toward local bodies for the encouragement of investigation and the diffusion of knowledge, it should act as an inspiring example and a reliable source of support. And in the broad field of international cooperation, it should unite with the leading academies of the world in the endeavor to perfect the organization of research and in the use of all agencies contributing to its advancement.

Needs of the Academy

Many of these objects have been accomplished by the National Academy in the past, but others remain for the future. The greatest aid in accomplishing its full work would be met by the provision of a suitable academy building, and an endowment sufficient to publish *Proceedings,* conduct research, provide public lectures, maintain exhibits illustrating current investigations, and to meet such additional needs as are implied by the Academy's national charter and its obligations to the scientific world and the general public. Through the courtesy of the Smithsonian Institution, extended in the year of the academy's organization, the annual meetings are held in the National Museum, in rooms ordinarily employed for other purposes. Thus the Academy does not even possess a permanent office, or a room for its library, which will be needed in the future for its work of research. It has therefore been compelled from the outset to decline many offers of books, and thus a large and valuable collection, comprising publications offered by many of the great academies, laboratories and observatories of the world, has been lost.[5]

It is difficult to overestimate the value of a suitable building in commanding public appreciation and support for any institution. Visible evidence of the Academy's existence is a matter of no small importance, when it is remembered that the average American citizen, though well-acquainted with the name of the Paris Academy through press reports of discoveries announced there, has never heard of our own national organization. But a building used as a storehouse and occupied but once a year is not enough. The Academy must be known as a living and active body, which recognizes and fulfills its many duties to science and the public. If its headquarters were constantly employed for such purposes as are enumerated later, the Academy would soon be looked upon as the natural source of information regarding the latest developments of science, and more generally recognized as the national representative of American research.

Importance of Publishing Proceedings

As explained in a previous paper, the name of the National Academy has never been associated with the work of its members, since the papers read at its meetings have not been published by the Academy. Thus it has not been sufficiently identified with the progress of American research, and the chief source of the reputation of the Paris Academy and the Royal Society has been lacking. But though the Academy would become more widely known by the publication of *Proceedings,* it would be foolish to take such a step merely to accomplish this purpose. The establishment of a new journal, in these days when the literature of science has become exceedingly complex, should never be undertaken without serious consideration of its probable usefulness. If it fulfills no good and lasting purpose, its life will be deservedly short. Hence we may not imitate the example of societies which establish their publications before the special journals had taken the field. We must recognize, on the one hand, that the various journals devoted to particular branches of science meet a clearly defined need and should not be rivaled, even to the apparent advantage of the Academy. On the other hand, we must also remember that the members of the Academy have adopted a regular plan of publication, the interruption of which might interfere with the accessibility of their papers. Thus, if *Proceedings* are to be established, they should be so planned as to

[5]The Academy has accepted some gifts of books, which are packed away (unbound) in the storerooms of the Smithsonian Institution.

serve a useful scientific end and be distinctly advantageous, not merely to the Academy itself, but to all of its members.

I am strongly of the opinion that no step which can be taken at the present time would be so beneficial to the National Academy as the publication of *Proceedings* containing the first announcements of important advances and the chief results of American research. I believe, furthermore, that this can be done in such a way as to benefit the members and contribute to the advancement of science. In many departments of the Academy's work papers published in the special American journals of limited foreign circulation do not reach a sufficiently large group of European readers. I am told that this is particularly true in biology, where American investigators are producing a great body of results of the first importance. Thus the *Proceedings* of the Academy, if properly distributed, might be made to serve the very useful purpose of bringing the work of a large number of investigators to the attention of scholars abroad. But in order to preserve all interests, and to interfere in the least degree with present plans of publication, the *Proceedings* should not be designed to occupy such a place as the special journals adequately fill.

[The chief advantage of the *Proceedings* would not be the same in all departments of science. In mathematics, where the existing journals are greatly overcrowded, prompt publication of the condensed results of new research would be heartily welcomed. The same thing is true in botany and in many other subjects. In fact, improved means of prompt publication would be generally appreciated by Academy members. In biology, as already remarked, the great number of special journals prevents many of them from reaching European laboratories, where American research is frequently overlooked as a consequence. In astronomy and astrophysics, which have fewer journals, the circulation of the chief American journals is large, and their contents reach all investigators abroad. But the practise of publishing separate series of circulars or bulletins, which has been adopted by many American observatories, confines the circulation of their papers to the limited number of astronomers and observatories on their mailing lists. If brief accounts of the broader aspects of these investigations were printed by the Academy, they would be useful to astronomers making a general survey of progress in their own field. But they would be even more serviceable to the mathematician, physicist, meteorologist, chemist, geologist or other investigator who may find information of direct or suggestive value in the results of astronomical research. Conversely, even those astronomers who keep in touch with progress in mathematics or physics can not also examine the numerous journals of chemistry, geology and other subjects which contain results applicable in their own work. It will thus be seen that the Academy could perform an important service in its special province of correlating knowledge by publishing papers covering the whole range of science.

The value of the *Proceedings* in strengthening the position of American science at home and abroad should not be overlooked. The rapid progress of American research in a single field may be known to the European specialist, but he may not realize that similar advances in other departments have raised American science to a new level. Recognition of this fact is desirable, not for the gratification of national pride, but because the international influence of America in science will grow with its prestige. The combination of effort which the *Proceedings* would represent, and the demonstration they would afford of American activity in research, are factors of real significance in securing that recognition and standing, both at home and abroad, which is needed to accelerate future progress.]

To accomplish the desired result, it would seem that the *Proceedings* should be intermediate in character between the *Comptes Rendus* of the Paris Academy and the *Proceedings* of the Royal Society. Papers read before the Paris Academy on Monday are printed and issued in the *Comptes Rendus* on the following Saturday—a record for speed which we should not expect to rival. Such accelerated publication, while it doubtless possesses certain advantages, renders impossible that more leisurely editorial examination which most journals demand. The *Proceedings* of the Royal Society, on the other hand, appear at irregular intervals, and frequently contain long and detailed papers, which with us might better find a place in the special journals. In the case of the National Academy it is doubtful whether publication at shorter intervals than one month is necessary, but the possible advantages of fortnightly publication should be carefully considered.

It goes without saying that papers for the *Proceedings*, while comparatively brief (perhaps averaging from three to five pages), should not be hasty announcements based on inadequate data. On the contrary, the dignity of the National Academy and the best interests of its members demand that only carefully matured conclusions, resulting from prolonged observational or theoretical research, should appear under the Academy's imprint. Measures and other exact data needed to establish these conclusions would be a necessary part of such papers, though long numerical tables, profuse illustrations, and detailed accounts of minor topics should be reserved for publication in the special journals, to which members would continue to contribute as before. The Academy *Proceedings* would thus serve for the first announcement of discoveries and of the more important contributions to research, illustrated by line cuts and occasional halftones in the text, when essential to clearness, but free from unnecessary detail and extensive numerical data. Non-members, as well as members, should be invited to contribute, with the understanding that their papers are to be presented by a member of the Academy, as in the case of the Paris Academy and the Royal Society.[6]

The constitution of the National Academy already provides for the issue of *Proceedings,* as well as *Memoirs* and *Annual Reports.* In fact, as explained in a previous paper, three volumes of *Proceedings* were published, though they did not contain papers presented to the Academy. There is therefore no need of any radical departure requiring amendment of the constitution. In other words, if sufficient funds are available, this very important step toward the development of the Academy can be taken by simple affirmative vote.[7]

The annual volumes of the *Proceedings,* bringing together for the first time the best product of American research, would place the Academy in a clearer light before the academic world. *Annual Reports* and infrequent volumes of *Memoirs* receive scant attention, except from a few specialists, in the libraries of our contemporary societies. But the *Proceedings,* published at regular intervals, and containing a standing notice of the Academy's publications, would aid in making them better known. The quarto *Memoirs,* eleven volumes of which have already appeared, afford an excellent place for extended publication, when the necessity for lengthy tables, numerous plates, or long discussions of data places the manuscript beyond the reach of the special journals. The publication of the *Proceedings* might serve to disclose much material worthy of use in the *Memoirs,* and the editorial board should be constantly on the watch for opportunities to extend the *Memoirs* and to render them more serviceable to science.

Science and the Public

The circulation of the *Proceedings* would necessarily be limited to scholars and scholarly institutions — they could not be expected to reach the general public. Here a difficulty remains to be overcome, since the results of original investigations should certainly be made more generally known and more clearly understood than they are at the present time. The average man of science, after sad experience with the daily press, is usually forced to the conclusion that newspaper publication is synonymous with rank sensationalism. Repeatedly told, and not without justice, that his cloistered wisdom should reach a wider world, he sometimes yields to the persistent demands of a reporter. The outcome is too well known to require telling. Even in the

[6]The *Proceedings* should be so planned as to interfere in the least possible degree with the *Journal* of the Washington Academy of Sciences, which is a publication similar in character to the one here proposed. As the *Journal* is devoted mainly to work done in Washington, or presented before the various Washington societies (other than the National Academy), no important overlapping of the two publications need be anticipated, especially as members of this Academy have rarely contributed to the *Journal*.
[7][The Academy voted, at its meeting of November, 1913, to begin the publication of *Proceedings* as soon as arrangements could be perfected. The first number will appear in January, 1915.]

case of a really intelligent and conscientious reporter, who does not distort or exaggerate, the "headline man" may be depended upon to provide a grotesque disguise. A few experiences of this sort suffice for most investigators. They are soon forced to shut out the reporter, and are well pleased when they succeed. Yet they recognize that the exclusion of the public from all contact with their work is neither fair nor desirable. Some way should be found of bridging the gap.

A plan followed in England by the Royal Society, of circulating brief abstracts on the day when a paper is read, which are afterwards published in *Nature* (sometimes in condensed form), is one which we might advantageously imitate. When a paper is accepted by the editorial board for publication in the *Proceedings*, a brief abstract, preferably prepared by the author, should be sent to *Science* (and perhaps also to *Nature*). At the same time this abstract, or a briefer one in less technical language, might be communicated to the Associated Press. It goes without saying that papers for the *Proceedings* would differ widely in their availability for popular treatment. Probably only a comparatively small proportion of them would contain results suitable for use by the Associated Press, but all would doubtless be published in abstract by *Science*. Through the Associated Press, and also through certain conservative newspapers and magazines, the Academy could thus bring before the public the actual results of scientific research, as distinguished from the false and distorted conceptions of science which most of our newspapers now disseminate.

Lectures on Research

The plan of publication outlined above is but one of several methods by which the Academy may enlarge its usefulness. Public lectures should also be instituted, primarily for the benefit of the Academy members, but also with the expectation of reaching a larger circle. Here the Academy would do well to study and imitate the Royal Institution of London, where original research and the diffusion of knowledge are combined in a very effective manner. In brilliant addresses, illustrated by lantern slides and experiments, a long line of illustrious speakers, best typified by Faraday, have charmed and enlightened the most distinguished audiences. Many of these speakers, including Davy, Faraday, Tyndall, Dewar, Rayleigh and Thomson, have been drawn from the staff of the Royal Institution. But their English contemporaries, as well as scientific men from all parts of Europe and the United States, have also been invited to describe their latest advances. The speaker at a "Friday Evening Discourse" is faced by the leaders of English thought and action in many fields. Privileged to select from the large collection of historic instruments accumulated during a century, and even to illustrate his points with the apparatus of Faraday himself, he feels an inspiration that no other platform affords. In such an atmosphere he learns to appreciate the dignity of popular science at its best, and to perceive how the busiest and most successful of present-day physicists can find time to deliver elaborate courses of Christmas lectures to a juvenile audience. These lectures, instituted by Faraday, are now in their eighty-seventh season. Under such topics as "The Chemistry of Flame" they have afforded him and his followers an opportunity to show how simply and beautifully the principles of science can be made to appeal even to young children.[8] The art of the popular lecture should be developed in the United States by the National Academy. Under its auspices, and with the example of the Royal Institution behind him, the lecturer need not fear for his dignity. The Academy would soon find its reward in the increasing appreciation of its work and purposes, the spread of scientific knowledge, and ultimately in larger endowments for research.

As a first step in this direction, the children of the late William Ellery Hale have established a course of lectures in memory of their father. Their object in doing so is twofold. In the first place, it is hoped that the lectures may add to the attractiveness of the Academy meetings, both to the members and the public. Again, it is believed that by a suitable choice of lecturers and topics, the inter-relationship of the various fields of research represented in the Academy, and the light thrown by the methods of investigation or of interpretation employed in one field upon those of another, may be illustrated in an effective way. Moreover, the lectures will afford an opportunity of testing whether the Academy may not further assist in increasing public appreciation of the cultural and the industrial value of science.

Science in Education

In the Academy of Plato and the Alexandrian Museum the functions of an academy and a university were united, and the work of instruction went hand in hand with the development of new knowledge. The growth of the modern university has now removed from national academies their former work of teaching a body of students, but their opportunity to exert a favorable influence on the educational methods of the nation remains. The Institute of France, as planned by Talleyrand and Condorcet,[9] was to control public instruction and offer courses to advanced students. This was not carried out, but an instance of the same sort is afforded by the Academy of Munich, which has charge of the public instruction of Bavaria.

There is no apparent reason why our own National Academy should have a formal connection with educational institutions. But in harmony with its purpose to advance knowledge in the United States, it should contribute toward the development of the science of education and take advantage of the possibility of increasing public appreciation of the educational value of science.

[8] The last course of Christmas Juvenile Lectures, on "Alchemy," "Atoms," "Light," "Clouds," "Meteorites" and "Frozen Worlds," was given by Sir James Dewar.
[9] See C. Hippeau, *L'Instruction publique en France pendant la Révolution* (Paris, Didier, 1881), pp. 115, 228.

In a presidential address which excited great public interest in England, Sir William Huggins emphasized before the Royal Society the importance of science in education.[10] We need not dwell upon his arguments regarding the value of scientific training in developing the power of accurate observation and the habit of correct and cautious reasoning. But a more neglected phase of science in education — its power of awakening and expanding the imaginative faculty — may be referred to in his own words:

> Surely the master-creations of poetry, music, sculpture and painting, alike in mystery and grandeur, can not surpass the natural epics and scenes of the heavens above and of the earth beneath, in their power of firing the imagination, which indeed has taken its most daring and enduring flights under the earlier and simpler conditions of human life, when men lived in closer contact with Nature, and in greater quiet, free from the deadening rush of modern society. Of supreme value is the exercise of the imagination, that lofty faculty of creating and weaving imagery in the mind, and of giving subjective reality to its own creations, which is the source of the initial impulses to human progress and development, to all inspiration in the arts, and to discovery in science.

Of all the teachings of science, the principle of evolution makes by far the strongest appeal to the imagination. Isolated phenomena, however remarkable, acquire a new meaning when seen in its light. Minute details of structure in animals or plants, slight differences of the relative intensity of lines in the spectra of stars, may become of intense interest even to the elementary student if explained as steps in a great process of development. But, after all that has been said and written since the time of Darwin, we fail to take full advantage of our opportunity. Properly presented, a picture of evolution in its broadest aspects would serve better than any other agency to stimulate the imagination, to awaken interest in science, and to demonstrate that its cultural value is in no wise inferior to that of the humanities. To the average student, even physics and chemistry are distinct branches of science, each occupied with its own problems. Astronomy, he knows, concerns itself with the heavenly bodies, botany with plants, zoology with animals. But if he studies these subjects at all, he almost invariably fails to realize their relationship, because no binding principle, like that of evolution, is brought prominently to his attention or, at the best, is restricted in its application to some single organic or inorganic field.

When Humboldt wrote "Cosmos" and Huxley lectured on "A Piece of Chalk" and other subjects, they showed what might be accomplished in picturing the problems of science in a broad way. The National Academy is better qualified than any other body in America to demonstrate what can be done in the same direction with the rich store of knowledge acquired since their time. A course of lectures on evolution, beginning with an account of the constitution of matter, the transformation of the elements, and the electron theory; picturing the heav-

[10]Sir William Huggins, *The Royal Society* (London, 1906), p. 109.

enly bodies and the structure of the universe, the evolution of stars and planets, and the origin of the earth; outlining the various stages of the earth's history, the formation and changes of its surface features, the beginning and development of plant and animal life; explaining modern biological problems, the study of variation and mutation, and the various theories of organic evolution; summarizing our knowledge of earliest man, his first differentiation from anthropoid ancestors, and the crude origins of civilization; and connecting with our own day by an account of early Oriental peoples, the rise of the Egyptian dynasties, and their influence on modern progress: such a course, free from technicalities and unnecessary details, richly illustrated by lantern slides and experiments, and woven together into a clear and homogeneous whole, would serve to give the average student a far broader view of evolution than he now obtains, and leave no doubt in the hearer's mind as to the cultural and imaginative value of science.

The William Ellery Hale lectures will open with a series on evolution, so designed as to be of interest to members of the academy, and at the same time to be intelligible and attractive to the public. At each meeting two lectures will be given by a distinguished European or American investigator, chosen because of his competence to deal with some branch of the subject. The first course of lectures, to be given by Sir Ernest Rutherford at the annual meeting in April, 1914, will deal with the constitution of matter and the evolution of the elements.[11] At the conclusion of this series, which will extend through several years, it is hoped that the lectures may be brought together, in a homogeneous and perhaps somewhat simplified form, into a small volume suitable for use in schools.

The course above outlined will serve to test the question whether the Academy may advantageously enter more extensively into the lecture field. So far as the members of the Academy are concerned, it seems probable that lectures by able American and European investigators would add to the interest of the meetings. But the value of the lectures to the general public can only be determined by experiment. If a suitable building can be obtained, and the success of these lectures is sufficient to warrant it, the foremost investigators, American and foreign, might be invited from time to time throughout the year to describe and illustrate their advances in the lecture-hall of the Academy. This plan is already followed by various American institutions, but the Academy, because of its national character, would be better able to attract the best men and to give their lectures more than local significance. Ample facilities for experimental illustration would also go far toward enhancing the value of the lectures. In short, the example of

---

[11][The second course was given at the autumn meeting by Dr. William Wallace Campbell on "Stellar Evolution and the Formation of the Earth."]

the Royal Institution should be followed as closely as possible.[12]

Industrial Research

The value of science to the American manufacturer, though no new theme, is capable of wide development at the hands of the National Academy. In a presidential address delivered before the Royal Society in 1902, Sir William Huggins dwelt on the "Supreme Importance of Science to the Industries of the Country, which can be secured only through making Science an Essential Part of all Education." He saw the fruits of English discoveries passing into the hands of Germany, whose universities have so long fostered and spread abroad the spirit of research, and wondered at the apathy of the average British manufacturer toward scientific methods. Huggins, speaking in plain language, pointed to the chief source of weakness — "the too close adherence of our older universities, and through them of our public schools, and all other schools in the country downward, to the traditional methods of teaching of medieval times."[13]

In this country, where the classics do not dominate the university system, the task of arousing an adequate appreciation of the enormous benefits which science can render is a far easier one. We must have, first of all, a widespread interest in science and some comprehension of its problems and methods. A general course on evolution, given to all college students, should be of great service as an entering wedge. More students might thus be led to take science courses, while those who specialize in the humanities could gain a better conception of what science means. The rapid development of research in our universities and technical schools promises to influence the faculties of our colleges, where a man's success as a teacher will be materially enhanced if he is also a producer of new knowledge. Thus the future is promising in the educational field.

On the side of our manufacturers, who are eager to adopt the most efficient methods, the outlook is equally favorable, as President Little of the American Chemical Society showed so effectively in his address on "Industrial Research in America."[14] Many great firms are establishing large research laboratories, where problems of all kinds are under investigation. The development within the past few years of Taylor's efficiency system is another indication that the advantages of scientific methods are being grasped and applied in the arts. But the opportunities in this direction are almost endless, and the National Academy would do well to devise ways and means of convincing not only the large manufacturers, but the small manufacturers as well, of the industrial importance of scientific research. Lectures on recent ad-

[12][It has been suggested by several members that these lectures might be repeated in two or three large cities, in cooperation with local scientific institutions.]
[13]Huggins, *The Royal Society*, p. 29.
[14]*Science, 38,* pp. 643-656, 1913.

vances in engineering, by European and American leaders, should have a powerful influence if carefully planned and effectively illustrated. Parsons on the steam turbine,[15] Marconi on wireless telegraphy,[15] Goethals on the Panama Canal, would attract large audiences and appeal in published form to a wide public.

But while the advantages resulting from ingenuity and invention and the best practise of engineering should certainly be brought out in the course of lectures I now have in mind, the improvement of manufactured products by research methods, and the potential industrial value of pure science are the points which should be emphasized. We have a long way to go before any single manufacturing firm employs seven hundred qualified chemists, as the combined chemical factories of Elberfeld, Ludwigshafen and Treptow do. The supremacy in this field of Germany, which produced chemicals valued at $3,750,000,000 in 1907, is directly due to the carefully directed research of an army of chemists, who learned the methods of investigation in the universities and technical schools.[16] The Berlin Academy of Sciences has also contributed in an important way to this result, through van't Hoff's investigations of the Stassfurth salt deposits. The recent rapid development of our own chemical industries leads us to hope that similar advances may soon be achieved in the United States. In electrical engineering, at least, we are already making comparable progress.

But the average man of business is much better able to appreciate the value of research directly applied to the improvement of manufactures than to comprehend the more fundamental importance of pure science. We must show how the investigations of Faraday, pursued for the pure love of truth and apparently of no commercial value, nevertheless laid the foundations of electrical engineering. If we can disseminate such knowledge, which is capable of the easiest demonstration and the most striking illustration, we can multiply the friends of pure science and secure new and larger endowments for physics, chemistry and other fundamental subjects.

[While there can be no doubt of the importance of emphasizing the value of industrial research, the necessity of vigilance in the interests of pure science is shown by the opposite tendency of several recent writers, who measure science solely in terms of its applicability in the arts.

The stimulus of commercial rivalry is doubtless a factor in the rapid progress of our great industrial laboratories, but I doubt if their di-

---

[15]Lectures before the Royal Institution, 1911.
[16]In 1910 the Nobel prize for chemistry went to Germany for the sixth time, thus giving to a single country sixty per cent. of all the Nobel prizes for chemistry awarded up to that date.

rectors would maintain that all chemical research should be of the industrial kind. Immediate commercial value as a criterion of success will not often point the way to the discovery of fundamental laws, though these are by far the richest source of ultimate achievement, practical as well as theoretical. Modern electrical engineers do not forget the investigations of Faraday and Hertz in pure science, nor do leading industrial chemists overlook the researches of Gibbs, van't Hoff, and others, which brought them no practical returns, but rendered many modern industries possible. Exclusive attention to industrial research means nothing more or less than the growth of the superstructure at the expense of the foundations. Industrial laboratories are able to offer large salaries and other tempting promises of material advantages, and thus to draw the most promising men from the universities. But while these laboratories should be strongly encouraged, and multiplied to the point where every small manufacturer will realize the value of research methods, this should not be done at the serious expense of pure science. Germany's success on the industrial side is primarily due to her still greater achievements in the university laboratories. The National Academy, by helping to maintain the two phases of American research in stable equilibrium, can perform a service which the truest advocates of applied science will recognize as essential to sound progress.]*

*This article was the third in the series and was followed by a fourth, "The Proceedings of the National Academy as a Medium of Publication," *Science, 41,* 815-817, 1915. The entire series was issued as a pamphlet, *National Academies and the Progress of Research* (Lancaster, Pa., 1915). [Editors]

The 200-inch Hale telescope looking northwest. Drawing by Russell W. Porter, 1939.
The Hale Observatories.

The Possibilities of Large Telescopes

Like buried treasures, the outposts of the universe have beckoned to the adventurous from immemorial times. Princes and potentates, political or industrial, equally with men of science, have felt the lure of the uncharted seas of space, and through their provision of instrumental means the sphere of exploration has rapidly widened. If the cost of gathering celestial treasure exceeds that of searching for the buried chests of a Morgan or a Flint, the expectation of rich return is surely greater and the route not less attractive. Long before the advent of the telescope, pharaohs and sultans, princes and caliphs built larger and larger observatories, one of them said to be comparable in height with the vaults of Santa Sophia. In later times kings of Spain and of France, of Denmark and of England took their turn, and more recently the initiative seems to have passed chiefly to American leaders of industry. Each expedition into remoter space has made new discoveries and brought back permanent additions to our knowledge of the heavens. The latest explorers have worked beyond the boundaries of the Milky Way in the realm of spiral "island universes," the first of which lies a million light-years from the earth while the farthest is immeasurably remote. As yet we can barely discern a few of the countless suns in the nearest of these spiral systems and begin to trace their resemblance with the stars in the coils of the Milky Way. While much progress has been made, the greatest possibilities still lie in the future.

I have had more than one chance to appreciate the enthusiasm of the layman for celestial exploration. Learning in August, 1892, that two discs of optical glass, large enough for a forty-inch telescope, were obtainable through Alvan Clark, I informed President Harper of the University of Chicago, and we jointly presented the opportunity to Mr. Charles T. Yerkes. He said he had dreamed since boyhood of the possibility of surpassing all existing telescopes, and at once authorized us to telegraph Clark to come and sign a contract for the lens. Later he provided for the telescope mounting and ultimately for the building of the Yerkes Observatory at Lake Geneva, Wisconsin.

In 1906 Mr. John D. Hooker of Los Angeles, a business man interested in astronomy, agreed to meet the cost of making the optical parts for an 84-inch reflecting telescope in the shops of the Mount Wilson Observatory in Pasadena, where a 60-inch mirror had recently been figured by Ritchey. Before the glass could be ordered he increased his gift to provide for a still larger mirror. Half a million dollars was still needed for the mounting and observatory building, and Mr. Carnegie, who was greatly taken with the project during his visit to the Observatory in 1910, wanted the Carnegie Institution of Washington to supply it. The entire income of the Institution was required, however, to provide for the annual expenses of its ten departments of research, of which the Observatory is one. Nearly a year later I was on my way to

Harper's Magazine, 15, 639–646, April 1928.

Egypt. At Ventimiglia, on the Italian frontier, I bought a local newspaper, in which an American cable had caught my eye. Mr. Andrew Carnegie, by a gift of ten million dollars, had doubled the endowment of the Carnegie Institution of Washington. A paragraph in his letter to the Trustees especially appealed to me: "I hope the work at Mount Wilson will be vigorously pushed, because I am so anxious to hear the expected results from it. I should like to be satisfied before I depart, that we are going to repay to the old land some part of the debt we owe them by revealing more clearly than ever to them the new heavens."

I hope that the 100-inch Hooker telescope, thus named at Mr. Carnegie's special request, has justified his expectations. Its results, described in part in *The New Heavens, The Depths of the Universe,* and *Beyond the Milky Way* have certainly surpassed our own forecasts. They have given us new means of determining stellar distances, a greatly clarified conception of the structure and scale of the Galaxy, the first measures of the diameter of stars, new light on the constitution of matter, new support for the Einstein theory, and scores of other advances. They have also made possible new researches beyond the boundaries of the Milky Way in the region of the spiral nebulae. Moreover, they have convinced us that a much larger telescope could be built and effectively used to extend the range of exploration farther into space. Lick, Yerkes, Hooker, and Carnegie have passed on, but the opportunity remains for some other donor to advance knowledge and to satisfy his own curiosity regarding the nature of the universe and the problems of its unexplored depths.

El Karakat, an Arabian astronomer who built a great observatory at Cairo in the twelfth century, once exclaimed to the Sultan, "How minute are our instruments in comparison with the celestial universe!" In his day the amount of light received from a star was merely that which entered the pupil of the eye, and large instruments were constructed, not with any idea of discovering new celestial objects, but in the hope of increasing the precision of measuring the positions of those already known. Galileo's telescope, which suddenly expanded the known stellar universe at the beginning of the seventeenth century, had a lens about 2¼ inches in diameter, with an area eighty times that of the pupil of the eye. This increase in light-collecting power was sufficient to reveal nearly half a million stars (over the entire heavens), as compared with the few thousands previously within range. The 100-inch mirror of the Hooker telescope, which collects about 160,000 times as much light as the eye, is capable of recording photographically more than a thousand million stars.

While the gain since Galileo's time seems enormous, the possibilities go far beyond. Starlight is falling on every square mile of the earth's surface, and the best we can do at present is to gather up and concentrate the rays that strike an area 100 inches in diameter. From an engineering standpoint our telescopes are small affairs in comparison with modern battleships and bridges. There has been no such increase in size since Lord Rosse's six-foot reflector, completed in 1845, as engineering advances would permit, though advantage has been taken of the possible gain in precision of workmanship. The time thus seems to be ripe for an examination of present opportunities, which must be considered in the light of recent experience.

I have never liked to predict the specific possibilities of large telescopes, but the present circumstances are so different from those of the past that less caution seems necessary. The astronomer's greatest obstacle is the turbulence of the earth's atmosphere, which envelops us like an immense ocean, agitated to its very depths. The crystal-clear nights of frosty winter, when celestial objects seem so bright, are usually the very worst for observation. Watch the excessive twinkling of the stars, and you will appreciate why this is true. In a perfectly quiet and homogeneous atmosphere there would be no twinkling, and star images would remain sharp and distinct even when greatly magnified. Mixed air of varying density means irregular refraction, which causes twinkling to the eye and boiling images, blurred and confused, in the telescope. Under such conditions a great telescope may be useless.

This is why Newton wrote in his *Opticks:*

If the Theory of making Telescopes could at length be fully brought into practice, yet there would be certain Bounds beyond which Telescopes could not perform. For the Air through which we look upon the Stars, is in a perpetual Tremor; as may be seen by the tremulous Motion of Shadows cast from high Towers, and by the twinkling of the fix'd stars. The only remedy is a most serene and quiet Air, such as may perhaps be found on the tops of the highest Mountains above the grosser Clouds.

Even at the best of sites, in a climate marked by long periods of great tranquillity, unbroken by storms, the atmosphere remains the chief obstacle. For this reason we could not be sure how well the 60-inch and 100-inch reflecting telescopes would work on Mount Wilson until we had rigorously tested them. Large lenses or mirrors, uniting in a single image rays which have traveled through widely separated paths, are more sensitive than small ones to atmospheric tremor. So it has always been a lottery, as we frankly told the donors of the instruments, whether the next increase in size might not fail to bring the advantages we sought.

Fortunately we have found, after several years of constant use, that on all good nights the gain of the 100-inch Hooker telescope over the 60-inch is fully in proportion to its greater aperture. The large mirror receives and concentrates in a sharply defined image nearly three times as much light as the smaller one, with consequent immense advantages. But the question remains whether we could now safely advance to an aperture of 200 inches, or, better still, to 25 feet.

Our affirmative opinion is based not merely upon the performance of the Hooker telescope, but also upon tests of the atmosphere made with apertures up to 20 feet. The Michelson stellar interferometer, with which Pease has succeeded in measuring the diameters of several stars, is attached to the upper end of the tube of the Hooker telescope. When its two outer mirrors are separated as far as possible, they unite in a single image beams of starlight entering in paths 20 feet apart. By comparing these images with those observed when the mirrors are 100 inches or less apart, Pease concludes that an increase of aperture to 20 feet or more would be perfectly safe. For the first time; therefore, we can make such an increase without the uncertainties that have been unavoidable in the past.

Other reasons that combine to assure the success of a larger telescope are the remarkable opportunities for new discoveries revealed by recent astronomical progress and the equally remarkable means of interpreting them afforded by recent advances in physics.

These new possibilities are so numerous that I must confine myself to three general examples, bearing upon the structure of the universe, the evolution of stars, and the constitution of matter. A 200-inch telescope would give us four times as much light as we now receive with the 100-inch, while a 300-inch telescope would give nine times as much. How would this help in dealing with these questions?

The first advantage that strikes one is the immense gain in penetrating power and the means thus afforded of exploring remote space. The spiral structure of nebulae beyond the Milky Way was unknown until Lord Rosse discovered it with his six-foot reflector in 1845. The Hooker telescope, greatly aided by optical and mechanical refinements and by the power of photography, can now record many thousands of these remarkable objects. Moreover, in the hands of Hubble it has shown that they are in fact "island universes," perhaps similar in structure to the Galaxy, of which our solar system is an infinitesimal part.

Our present instruments are thus powerful enough to give us this imposing picture of a universe dotted with isolated systems, some of them probably containing millions of stars brighter than our sun. It is also possible to measure the distance of the Great Nebula in Andromeda and one or two other spirals that lie about a million light-years from the earth. Much larger telescopes are needed, however, to continue the analysis of these nearest spirals, now only just begun, and to extend it to some of those at greater distances. Needless to say, the greater power of larger telescopes would also give us a far better understanding than we now possess of the structure and nature of the Galaxy, of which we still have much to learn. For example, we cannot yet say whether it shares the characteristic form of the spiral nebulae, nor do we even know with certainty whether it rotates about its center at the enormous velocity that seems equally characteristic of the "island universes." In fact, our own stellar system offers countless opportunities for productive research, as the important advances in our knowledge of the Galaxy recently made by Seares with the 60-inch Mount Wilson reflector so clearly indicate.

If our ideas of the structure of the universe are thus in a very early stage, the same may be said of our knowledge of the evolution of the stars. Recent discoveries in physics have greatly modified our conception of stellar evolution, affording a rational explanation of scores of questions formerly unanswered, but raising many new and fascinating problems. Giant stars with diameters several hundreds of times that of the sun, expanded by internal pressure to gossamer tenuity, lie near one end of our present stellar vista, with dwarfs of a density more than fifty thousand times that of water near the other. The sun, a condensing dwarf, 1.4 times as dense as water, stands on the downward slope of stellar life. The continual radiation that marks the transition from giant to dwarf is now attributed to the transformation of stellar mass into radiant energy, thus harmonizing with Einstein's views and accounting for the decrease in mass observed with advancing age. Surface temperatures ranging from about 3000° C. in the earlier stage of stellar life to about 100,000° at its climax, and internal temperatures perhaps reaching hundreds of millions of degrees are among the incidents of stellar existence. But here again, while theory and observation have recently joined in painting a new and surprising picture of celestial progress, important differences of opinion still exist and many of these await a more powerful telescope to discriminate between them. For while theories based on modern physics have been our chief guide in recent years, the final test is that of observation, and often our present instruments are insufficient to meet the demand.

So much in brief for the questions of celestial structure and evolution, though I have had to pass over the greatest of these problems: that of determining with certainty the successive stages in the development of the spiral nebulae, a phase of evolution vastly transcending that involved in the birth, life, and decline of a particular star. I have space to add only a word regarding the role of great telescopes in the study of the constitution of matter.

The range of mass, temperature, and density in the stars and nebulae is of course incomparably greater than the physicist can match in the laboratory. It is, therefore, not surprising that some of the most fundamental problems of modern physics have been answered by an appeal to experiments performed for us in these cosmic laboratories. For example, one of the most illuminating tests of Bohr's theory of the atom has just been made at the Norman Bridge Laboratory by Bowen in a study of the characteristic spectrum of the nebulae, where the extreme tenuity of the gas permits hydrogen and nitrogen to exist in a state harmonizing with the theory but unapproachable in any vacuum-tube. Similarly, Adams' observations of the companion of Sirius with the Hooker telescope confirmed Eddington's prediction that matter can exist thousands of times denser than any terrestrial substance. In fact, things have reached such a point that a far-sighted industrial leader, whose success may depend in the long run on a complete knowledge of the nature of matter and its transformations, would hardly be willing to be limited by the feeble range of terrestrial furnaces. I can easily conceive of such a man adding a great telescope to the equipment of a laboratory for industrial research if the information he needed could not be obtained from existing observatories.

The development of new methods and instruments of research is one of the most effective means of advancing science. In hundreds of cases the utilization of some obvious principle, long known but completely neglected, has suddenly multiplied the possibilities of the investigator by opening new highways into previously inaccessible territory. The telescope, the microscope, and the spectroscope are perhaps the most striking illustrations of this fact, but new devices are constantly being found, and the result has been a complete transformation of the astronomical observatory.

From our present point of view the chief question is the bearing of these developments on the design of telescopes. To Galileo a telescope was a slender tube, three or four feet in length, with a convex lens at one end for an object glass, and a concave lens at the other for an eyepiece. With this "optic glass" the surprising discoveries described in the *Sidereus Nuncius* were made, which shifted the sun from its traditional position as a satellite of the earth to the center of the solar system, and greatly enlarged the scale of the universe. After his time the telescope grew longer and longer, finally reaching the ungainly form of a lens supported on a pole as much as two or three hundred feet from the eyepiece. The invention of the achromatic lens brought the telescope back to manageable dimensions and permitted the use of an equatorial mounting, equipped with driving-clock to keep the celestial object at rest in the field of view. With the improvement of optical glass the aperture steadily increased, finally reaching 36 inches in the Lick and 40 inches in the Yerkes telescope.

Meanwhile it had become clear that the reflecting telescope, designed by Newton to avoid the defects of single lenses, possessed many advantages over the refractor. Chief among these are its power of concentrating light of all colors at the same focus and the fact that the light does not pass through the mirror, but is reflected from its concave front surface. Speculum metal, a highly polished alloy of tin and copper, was used for the early reflectors, reaching a maximum size in Lord Rosse's six-foot telescope. Mirrors of glass, silvered on the front surface, were then introduced, and proved greatly superior in lightness and reflecting power. Moreover, optical glass perfect enough for lenses cannot be obtained in very large sizes, and even it if could, the loss of light by absorption in transmission through the glass would prevent its use for objectives materially exceeding that of the Yerkes telescope. Therefore, our hopes for the future must lie in some form of reflector.

It is evident that a lens, through which the starlight passes to the eye, must be mounted in a very different way from a concave mirror, which receives the light on its surface and reflects it back to the focus. The large concave mirror lies at the bottom of the telescope tube, which is usually of light skeleton construction, open at the top. The surface of the mirror is figured to a paraboloidal form, which differs somewhat from a sphere in curvature, and has the power of concentrating the parallel rays from a star in a point at the focus. This focus is near the top of the tube, opposite the center of the mirror.

For some classes of work it is desirable to place the photographic plate, small spectroscope, or other accessory instrument at this principal focus, centrally within the tube. Some starlight is thus cut off from the large mirror, but the loss is small and is less than with other arrangements. Newton interposed a plane mirror, fixed at an angle of 45°, which reflected the light to the side of the tube, where he placed the eyepiece. Cassegrain substituted a convex mirror for Newton's plane. Supported centrally at right angles to the beam, it changes the convergence of the rays and brings them to a focus near the large mirror. An inclined plane mirror may be used to intercept them, thus bringing the secondary focus at the side of the tube, or the large mirror may be pierced with a hole, allowing the rays to come to a focus close behind it.

In a third arrangement, the rays may be sent through the hollow polar axis of the telescope to a secondary focus at a fixed point in a constant temperature laboratory. This arrangement, first suggested by Ranyard, was embodied with both the Newtonian and Cassegrain methods in the mountings of the 60-inch and 100-inch telescopes of the Mount Wilson Observatory. By these means we may obtain any desired equivalent focal length (which varies with the curvature and position of the small convex mirrors) and thus photograph celestial objects on a large or small scale, as required by the problem in hand. Furthermore, we can use to the best advantage all types of spectroscope, photometer, interferometer, thermocouple, radiometer, photo-electric cell, and the many other accessories developed in recent years.

These accessory instruments and devices have made possible most of the discoveries of modern astrophysics. The stellar spectroscope, originally merely a small laboratory instrument attached to a telescope, has grown to the dimensions of the powerful fixed spectrograph of 6 inches aperture and 15 feet in length, recently used with splendid success by Adams in photographing the spectra of some of the brightest stars. The development of this method of high dispersion stellar spectroscopy, initiated in the early days of the Yerkes Observatory, was one of my chief incentives in endeavoring to obtain large reflecting telescopes for the Mount Wilson Observatory. The recent advances in our knowledge of the atom and the consequent complete transformation of spectroscopy from an empirical to a rational basis greatly increase the possibilities of analyzing starlight. In most of the small-scale spectra photographed with ordinary stellar spectrographs the lines are so closely crowded together that they cannot be separately measured. With a larger telescope we could push the dispersion to the point attained by Rowland in his classic studies of the solar spectrum, and thus take full advantage of the great possibilities of discovery offered us by recent advances in physics.

These details are important because they point directly to the type of telescope required. It is true that in some cases lenses may be used instead of convex mirrors for enlarging the image; but in our judgment the design should permit observations to be made in the principal focus of the large mirror, at a secondary focus just below the (pierced) mirror, and at another secondary focus in a fixed laboratory. It should also be possible to attach to the tube a large Michelson stellar interferometer, arranged for rotation in position angle and thus suitable for the measurement of very close double stars.

A mounting designed by Pease of the Mount Wilson Observatory meets these requirements and is worthy of careful consideration. It is large enough to carry a mirror 25 feet in diameter, collecting nine times as much light as the 100-inch Hooker telescope. It would thus enlarge our sphere of observation to three times its present diameter and increase the total number of galactic stars to three or four times that now within range.

This, of course, is a tentative design, subject to modification in the light of an exhaustive study. Of all the optical and mechanical problems involved only one presents real difficulties, but there is no reason to think that these cannot be readily surmounted. This is the manufacture of the glass for the large mirror.

Our chief difficulty in the case of the Hooker telescope was to obtain a suitable glass disc. The largest previously cast was that for the 60-inch mirror of our first large reflector. This is 8 inches thick and weighs a ton. The 100-inch disc, 13 inches thick, weighs nearly five tons. To make it three pots of glass were poured in quick succession into the mold. After a long annealing process, to prevent the internal strains that result from rapid cooling, the glass was delivered to us. Unlike the discs previously sent by the French makers, it contained sheets of bubbles, doubtless due in part to the use of the three pots of glass, while but one had sufficed before. Any considerable lack of homogeneity would result in unequal expansion or contraction under temperature changes, and experiments were, therefore, continued at the glass factory in the Forest of St. Gobain in the hope of producing a flawless disc. As they did not succeed, the disc containing the bubbles was given a spherical figure and tested optically under a wide range of temperature. Its performance convinced us that the disc could safely be given a paraboloidal figure for use in the telescope, where it has served ever since for a great variety of visual and photographic observations.

Recently, important advances have been made in the art of glass manufacture, and mirror discs much larger and better than the 100-inch can now undoubtedly be cast. Pyrex glass, so useful in the kitchen and the chemical laboratory because it is not easily cracked by heat, is also very advantageous for telescope mirrors. Observations must always be made through the widely opened shutter of the dome, at temperatures as nearly as possible the same as that of the outer air. As the temperature rises or falls the mirror must respond. The small expansion or contraction of Pyrex glass means that mirrors made of it undergo less change of figure and, therefore, give more sharply defined star images — a vitally important matter in all classes of work, especially in the study of the extremely faint stars in the spiral nebulae, for which Pease's design is especially adapted.

Dr. Arthur L. Day of the Carnegie Institution of Washington, working in association with the Corning Glass Company, has succeeded in producing glass with a higher silica content than Pyrex and, therefore, with a lower coefficient of expansion. Moreover, Dr. Elihu Thomson and Mr. Edward R. Berry of the General Electric Company have recently made discs up to 12 inches in diameter of transparent fused quartz (pure silica), which is superior to all other substances for telescope mirrors. The chief difficulty in the manufacture of fused quartz has been the elimination of bubbles. These would do no harm whatever within a large telescope mirror, provided its upper surface were freed from them by a method proposed by Dr. Thomson. In fact, the presence of a great number of bubbles would be a distinct advantage in reducing the weight of the disc. As there is every reason to believe that a suitable Pyrex or quartz disc could be successfully cast and annealed, and as the optical and engineering problems of figuring, mounting, and housing it present no serious difficulties, I believe that a 200-inch or even a 300-inch telescope could now be built and used to the great advantage of astronomy.

Limitations of space have prevented mention of many interesting matters of detail. It goes without saying that all questions relating to the optical as well as the engineering design should be thoroughly investigated by a group of competent authorities, who should also include those best qualified to deal with related problems involving the design of spectroscopes and the many other accessory instruments required. As for photographic plates, it is well known that the power of photographic telescopes could be materially increased by improving their quality, so that no effort in this direction should be spared.

Perhaps a word as to procedure may be added. The first step should be to determine by experiment how large a mirror disc, preferably of fused quartz, can be successfully cast and annealed. Meanwhile all questions as to the final design of the mounting and accessories could be settled. With the completion of the mirror disc the only uncertainty would vanish and the optical and mechanical work could begin.

George Ellery Hale in the National Academy of Sciences.
Photograph by James Stokeley

3
Perspectives

Introduction

The legacy of George Ellery Hale is evident today in all the fields in which he worked. The following papers review some of these developments in the construction of large astronomical telescopes and related instrumentation, research in solar physics, and the role of scientific organizations. Several of the authors have played key roles in these fields and can be thought of as Hale's scientific heirs. The papers were originally presented at the Hale Centennial Symposium at the 1968 annual meeting of the American Association for the Advancement of Science.

C. Donald Shane, who first used the large Lick Observatory telescope more than fifty years ago, was director from 1945 to 1958. He played a leading role in the development of the 120-inch Lick reflector and in cooperative projects to build new astronomical facilities. His essay discusses the evolution of large astronomical telescopes during the past seventy-five years.

Ira S. Bowen was involved in the planning of the 200-inch Hale telescope and related optical systems in the mid-1930s and was the director of the Mount Wilson and Palomar Observatories (now the Hale Observatories) from 1946 to 1964, a period during which Hale's dream of the 200-inch became a reality and produced the results he had prophesied in 1928. Bowen reviews the development of the auxiliary instrumentation that has made the large telescopes such versatile and powerful tools for research.

Robert Howard's research at the Hale Observatories is concerned with magnetic fields of the sun and sunspots, the field in which Hale made pioneering contributions. Howard's review of the field traces its course from the time of Hale's initial efforts to today's computer-assisted research.

Daniel J. Kevles is a historian of science at the California Institute of Technology. He is concerned with the social history of American science in the twentieth century and evaluates Hale's vision of the role of private, academic, and governmental institutions in advancing science.

An aerial view of the Lick Observatory on Mount Hamilton with the dome of the 120-inch telescope in the foreground.
Lick Observatory.

Astronomical Telescopes since 1890

C. Donald Shane

George Ellery Hale probably contributed more to the general advance of astronomy than any other scientist of his time. His own astronomical research was enormously fruitful, but his greatest contribution lay in the conception and promotion of ever more powerful instruments of observation. With these he developed programs of research designed to penetrate the mysteries of the universe. He had the cooperation of an extraordinary staff, gathered and held through his rare qualities of leadership. His enthusiasm, backed by sound scientific judgment, encouraged sympathetic support for his increasingly larger instruments. These instruments grew from a 4-inch Clark telescope, somewhat reluctantly funded by his father, to his last great adventure, the 200-inch Hale telescope on Palomar Mountain. It may be said that the 200-inch, aside from mere size, represents the greatest combination of successful innovations in the history of telescope building. The fact becomes impressive with the realization that the failure of any one of these innovations could have been disastrous to the whole project. The result speaks highly for the courage and skill of those who participated in designing and building the Hale telescope.

Hale's telescopes were the culmination of a long tradition dating back to the first refracting telescope, which was built about 1610. Only a few years later a Jesuit priest built what was probably the first reflector. By the middle of the nineteenth century both refractors and reflectors had come a long way, and a lively controversy was developing between the advocates of each type of telescope. In 1877, Sir Howard Grubb, the English telescope maker, wrote: "A veteran and well-known worker with Refractors declared 'he never looked into a Reflector without drawing away his eye in disgust,' while workers with Reflectors cannot understand how the Refractor workers can bear that dreadful fringe of colour from the secondary spectrum."[1] Stated more objectively, one great advantage of a reflecting telescope is freedom from chromatic aberration. Another is the larger feasible size of a mirror, with the consequent increase in light-gathering power. A disadvantage is that the mirror surface finish must be perfected more than that of a lens if it is to achieve the same optical quality as a lens. Furthermore, for telescopes under 40 inches, the maximum practical diameter for a lens, a mirror is harder to support free of troublesome flexure.

The controversy between advocates of reflectors and refractors became more than a war of words when plans for the Lick telescope became a reality in 1876. In his First Deed of Trust, James Lick had directed that a telescope be built "superior to and more powerful than any telescope ever yet made...."[2] Captain Richard S. Floyd, chairman of the Lick trustees, sought advice from the most renowned astronomers in the world. In a long letter to Captain Floyd, British astronomer Sir David Gill summed up the situation: "Thus it is, and very properly has been the object of Regular observatories to possess a good Refractor, and it has been left to men of special mechanical genius, Herschel, Rosse, Lassell, Draper, Nasmyth etc. to make and employ large Reflectors.... It is also clear that the demands of science and the promise of the richest harvest of results point to the Reflector."[3] When the day of decision came, the trustees probably very wisely chose a 36-inch refractor as the major instrument for the new Lick Observatory on Mount Hamilton near San Jose, California, and it was mounted in 1888.

The large Lick telescope was surpassed in size in 1897 by the 40-inch refractor at the Yerkes Observatory at Williams Bay, Wisconsin. Alvan Clark & Sons, telescope makers of Cambridgeport, Massachusetts, designed and produced the optical components, drawing on their experience with the Lick instrument. These remain the world's largest refractors, and both are excellent instruments. To attempt a larger refractor of this type appears unrewarding. All its inherent disadvantages increase with size: the absorption of light in passing through the lens, the chromatic aberration, and the apparently unavoidable flexure of very large lenses. On the other hand, the development of refined skills for the finishing of large mirrors and the application of modern engineering techniques to overcome problems of support and mounting have opened the way to the successful development of ever larger reflectors. With the introduction of the photographic dry plate during the second half of the nineteenth century, it was apparent that reflectors, with their superior light-gathering power, had a further advantage over refractors.

Thomas Robinson, an Irish astronomer, wrote prophetically in 1876: "I repeat my belief that it is only by large Reflectors that we shall reach the utmost depths of celestial exploration in stellar spectroscopy, and the study of the remoter planets and Nebulae; and, instead of unwisely depreciating this power, it is a wiser course to consider how their real defects may be remedied."[4] The last ninety years have been largely directed toward this end. The field is far too extensive to include all its ramifications, so my discussion will essentially be limited to ground-based optical telescopes in which the principal advances have been made.

In 1885, Sir Howard Grubb built a 20-inch silver-on-glass reflector with an attached guiding telescope for Isaac Roberts's observatory in Sussex, England. It was especially designed for stability and accuracy of drive to permit taking long exposures. Imperfect by present standards, it still produced photographs of nebulae and star fields, by far the best of their time. Ten years later, the Lick Observatory acquired a 36-inch reflector donated by Edward Crossley of Halifax, England. The telescope had an excellent mirror made by George Calver, which is still in use, but its mounting was inadequate for the demands of stellar photography. When James Edward Keeler came to the Lick Observatory as director in 1898, he determined to master the idiosyncracies of the Crossley. In Keeler's skilled hands, as a commentator in 1908 noted, "This form of telescope was in effect born again."[5] Keeler died only two years after he became director, but in that short period he produced the finest photographs of nebulae that had ever been taken. He showed that at least 120,000 nebulae, a large fraction of them spirals, could be photographed with the Crossley.

The present Crossley mounting was designed by Charles D. Perrine about 1904. The new mounting enabled the telescope to be used to the full capabilities of the mirror, but it is awkward by modern standards. It is true that on a few occasions an astronomer fell from the platform to the floor ten or fifteen feet below. One observer commented wryly that the only comfortable and safe way to work would be to flood the dome with water and observe from a boat.

In the late 1890s, George Ritchey completed a 24-inch reflector at the Yerkes Observatory. In 1908 the 60-inch reflector and in 1917 the 100-inch were completed at Mount Wilson; and finally the 200-inch at Palomar Mountain was completed in 1949. These represented not only forward steps in size but also in sophistication of design. Much of this success was due to the imagination and organizational genius of Hale. Except for a 236-inch with an altazimuth mounting under construction in the Soviet Union, no larger reflectors are contemplated. However, improvements in design and construction continue at a rapid pace in connection with the several telescopes in the 150-inch bracket that are now under way.

A reflecting telescope is a complex instrument that can operate successfully only when its separate parts work effectively. The mirror is the heart of the telescope. There is first the preparation of the mirror base. Its surface must be figured to optical precision, and must then be coated with a reflecting material. When and if these difficult processes are completed successfully, the mirror must be supported in a cell so as to avoid optical distortion with changing positions of the telescope. Mounting a mirror in a large telescope is a major engineering challenge. Then there are the complex and exacting requirements of the telescope drives and controls. Finally, the telescope must be housed. Let us take a closer look at each stage in the process.

Early mirrors in general were made of speculum metal, usually various alloys of tin and copper. In 1875, Grubb built the "Great Melbourne Reflector" with a 4-foot speculum. Grubb noted that it had been remarked that "Reflectors very seldom do good work except in the hands of their makers."[6] This proved to be true in the case of the Melbourne reflector. When the mirror became tarnished, attempts to resurface it in Australia were unsuccessful, and the telescope was a sad disappointment and a temporary setback for the reflector cause. When interest in reflectors revived, metal mirrors were abandoned in favor of silver-on-glass on which the reflecting coat could be replaced without disturbing the optical surface.

A major problem has been to find materials that provide minimum distortion with changing temperature. The difficulty is more acute with larger mirrors, for these achieve temperature equilibrium more slowly. This was especially serious with the 100-inch plate glass disc at Mount Wilson. The development of low-expansion Pyrex glass by the Corning Glass Company resulted in great improvements in mirrors, because it had only one-fifth the temperature coefficient of plate. Large mirrors of this material were first made in connection with the Hale 200-inch telescope. A further innovation was the use of ribbed structure. Such mirrors attained temperature equilibrium more quickly, were lighter, and were adapted to better means of support. One of these mirrors originally cast for the Palomar program was used for the Lick Observatory 120-inch telescope, which was completed in 1959. This type of design for glass mirrors has become standard practice. An exception is the Isaac Newton telescope built at the Greenwich Observatory in England in 1968, which has a solid Pyrex disc.

At the time the Hale telescope was being designed in the 1930s, attempts were made to produce large fused silica discs. These had the advantage of a lower coefficient of expansion, about one-fourth to one-fifth that of low-expansion Pyrex. They also had a 20 percent higher conductivity, which speeded achievement of temperature equilibrium. There was, however, the disadvantage that they could not be cast in a ribbed structure. At that time, experiments in casting very large fused silica discs were not successful. Since then, the method has been perfected and an ultralow-expansion fused silica has become available. A new type of material, ceramic glass, is now being produced which has a practically zero coefficient of expansion at operating temperatures. The largest telescope mirrors currently under construction are made of fused silica or of ceramic glass.

Given time, things often come full circle. At the Kitt Peak National Observatory, they are again making metal mirrors. A 60-inch aluminum mirror forms the image in the McMath Solar Telescope. Such mirrors can readily be cast in a variety of structural forms. The large aluminum mirrors in use at Kitt Peak are first figured approximately, then heavily coated with Kanigen, a nickel product deposited chemically. The final figuring is done on the Kanigen surface. Insofar as time has permitted a test, these metal mirrors seem to have good stability against aging. The higher temperature coefficient of expansion is largely compensated by the greater heat conductivity. The problems of temperature distortion in mirrors seem to have been solved.

The art of figuring mirrors has kept up quite well with the astronomical requirements. The tradition of optical skill set by such artists as George Calver, Sir Howard Grubb, Alvan Clark & Sons, John Brashear, and George Ritchey has been carried on in modern times by a host of able successors. Modern telescopes in general call for short focal ratios and consequently steep mirror curvatures that are difficult to figure and test. Actually, at the earth's surface, the optical disturbance of the atmosphere limits the definition. The highest degree of perfection in figure is of no special benefit. However, it is now within the range of possibility to construct a large mirror whose defining performance is limited only by the size of the diffraction disc. When telescopes are used outside the earth's atmosphere, this will be an enormous advantage. Already, Martin Schwarzschild has used a 36-inch reflector supported by a balloon to make observations far out in the stratosphere.

The final step is to apply a reflecting coat on the figured mirror. Silver deposited chemically on glass was used almost universally until 1930, when the evaporation process of depositing metal coatings was introduced by John Strong and others. Since that time, reflecting surfaces of aluminum have been applied by evaporation, and in recent years the method has become routinely reliable due to improved vacuum techniques. The aluminum preserves its reflecting surface better than silver and has superior reflectivity in the more refrangible spectral regions. Even on the Kanigen surface this final aluminum coat can be removed and renewed without harm to the optical surface.

A major problem with large reflectors has been to devise an effective support system for the mirror. It must be supported both radially and axially, so that in its different orientations it will not distort harmfully under gravity. This requirement has been approached in a variety of ways. In 1869, T. R. Robinson and Thomas Grubb described earlier approaches: "In the first reflecting telescopes the specula were held by three front stops, against which they were kept by three or more screws behind. The screws were afterwards replaced by three springs, and this defective system remained until the Herschels substituted one unobjectionable in principle. They placed the speculum in a strong metal box resting on several folds of soft and elastic cloth."[7] Most small mirrors were supported by three back pads strategically placed. For larger mirrors, six pads were used, balanced about three centers. The edge or radial support was generally furnished by a metal band lined with some soft material and tightened as much as possible without introducing mirror distortion. The mirror of the 36-inch Crossley reflector is supported in this way. For still larger mirrors, such as the 60-inch at Mount Wilson, a system of counterweight levers was used. This was by no means a new idea in 1908, for Thomas Grubb had experimented with it thirty years earlier. In 1877 Grubb's son described this innovation and commented: "No matter what the size of the mirror may be, there is neither theoretically nor practically any difficulty in supporting it perfectly free from flexure by using my father's system of levers."[8] Counterbalanced pads were also used for the 100-inch at Mount Wilson. When friction developed between the back pads and the mirror, it was relieved by giving the pads two degrees of lateral freedom. The ribbed mirror of the 200-inch at Palomar provided a real breakthrough because the support holes permitted the use of counterbalanced radial supports distributed over the whole mirror. This has proved to be an excellent system and is standard for all ribbed discs.

The growing use of large solid mirrors requires a different approach. A few such mirrors are now supported against the back by a cushion of air whose pressure is adjusted according to the zenith distance of pointing. The solid disc of the 98-inch Isaac Newton telescope at Herstmonceux has radial supports around the edge with a counterbalanced lever system that pulls from above and pushes from below to prevent the disc from slumping. The axial support is provided by an air cushion. The 140-inch mirror of the European Southern Observatory on La Silla, Chile, is, to the best of my knowledge, unique in having mechanical axial supports and an air-cushion radial support, the reverse of the Isaac Newton telescope. The 61-inch fused silica mirror in the astrometric telescope of the U.S. Naval Observatory in Flagstaff has a pneumatic axial support, and the 35-inch secondary, which operates face down, is likewise held from the back by reduced air pressure. The radial support of the main mirror consists of a mercury-filled Neoprene tube that virtually floats the vertical component of the weight. The mirror is, of course, positioned by fixed supports that carry minimum loads.

The idea of air cushioning is not new. In the nineteenth century, Foucault supported his specula on air cushions that were inflated as required.[9] Sir Howard Grubb extended the idea to refractors in 1877, suggesting that "a far more elegant contrivance would be to hermetically seal the tube and fill the tube with air under pressure as would support a sufficient portion of the weight of the objective on a perfect air cushion. Of course this pressure should be regulated according to the altitude of the telescope, but I have devised a mechanical contrivance for this purpose."[10] So far, this plan for supporting large lenses has not been put into practice.

The most spectacular recent developments have been in the mounting of large reflectors. Until the 1930s, the emphasis was on trying to build tubes to combine lightness and rigidity. In the seventeenth century, Christian Huygens had solved the problem by eliminating the tube entirely. He mounted his lens on a mast and controlled it from the eyepiece, 123 feet distant, by means of a string. Two centuries later, astronomers were still working on the same problem. Most telescopes were light and open structures, strongly cross-braced. A few, including the remounted Crossley, had heavy solid tubes. It was, of course, not possible to make a telescope tube completely stiff. As a result, the ends sagged below the declination axis. Thus, the two ends of the tube were not parallel, and this gave a collimation displacement depending upon zenith distance. A remarkable advance was made by Mark Serrurier in the design of the 200-inch Hale telescope. He invented the now well-known structure that permits an equal amount of sag at the two ends of the tube and at the same time holds them parallel. Thus, for practical purposes the tube has zero flexibility. This type of construction is favored at present for most large telescopes. The problems of mounting have been considerably lessened by the perfection of short-focus telescopes. A focal ratio of f/5 was commonly in use around 1930. It was reduced to a focal ratio of 3.3 in the Hale telescope, and to 2.8 in the two 150-inch telescopes now being built for Kitt Peak and Cerro Tololo. This means shorter tubes, lighter mountings, small domes — all great structural advantages.

Large telescope mountings can be divided into two general categories, asymmetric and symmetric. In the asymmetric type, the telescope tube is placed to one side of the polar axle. Warner and Swasey used this mounting for the 72-inch at Victoria and recently for the 107-inch telescope at the McDonald Observatory. The firm of Grubb Parsons has used the same type of mounting for a series of telescopes in the 70- to 80-inch range. The first important modern telescope with the symmetric mounting was the Mount Wilson 60-inch. Most of the weight of this telescope is floated on mercury to reduce friction and to provide greater accuracy in the drive. This was not new, since in a primitive form it had been tried unsuccessfully in the original Crossley mounting. It appeared that the Mount Wilson 100-inch would be too massive for a fork mounting, and it was mounted in a yoke, again using mercury flotation. The yoke mounting has the disadvantage that a substantial area around the pole cannot be observed. Again inventive genius came to the rescue, this time with the oil pad bearings. Their use made possible the horseshoe yoke for the 200-inch, a mounting that combined the advantages of the fork and the yoke. Since 1949, when the 200-inch telescope was completed, the Lick Observatory 120-inch reflector is the largest telescope to have been built. It has a fork mounting. It owes a great deal, as do many others, to the experience with the 200-inch. The large telescopes now under construction use variations of the fork mounting.

The successful operation of a large telescope requires a complex assembly of telescope drives and controls. At one time, virtually everything about a telescope was powered by hand. But even in the early days, astronomers looked forward to an easier life. As Sir Howard Grubb wrote:

I am and always have been a strong advocate for making the observer as comfortable as possible (believing that thereby his capacity for useful work will be increased). I would strongly advocate in very large telescopes that hydraulic power be utilized for conducting all the laborious operations, so that the observer, without moving from his chair, might, simply by pressing one or the other of a few electric buttons, cause the telescope to move around in right ascension and declination, the dome to revolve, the shutters to open and the clock to be wound. This is no mere Utopian dream.... It is only in the application that there would be any difficulty encountered.[11]

It did not happen overnight, but his hopes of 1877 are now far surpassed.

Until about the 1920s and later, many telescopes were moved in hour angle by clocks wound by hand and driven by falling weights. Then synchronous motors took over with frequencies controlled by vibrating wires or other means. Automatic guiding by photoelectric devices has been introduced, not only relieving the astronomers of much drudgery but doing a better job than possible with the human hand and eye. Fully automated telescopes, at present in the experimental stage, provide for taping of the observing program and automatic monitoring of the observing conditions. The observations are reduced almost simultaneously in a computer which then makes adjustments in the program. The largest automatic telescope in the United States is one of 50-inch aperture at Kitt Peak, originally planned to give experience that would be valuable in the remote control of space telescopes. We have come a long way from Lassell's 48-inch reflector of 1861 in which the tracking was done by an assistant who turned a crank timed to the ticking of a clock.

Finally, we return to the question of housing telescopes. Many early telescopes were operated in the open because of the expense and the technical difficulty of providing shelter. During the past century the rotating dome has come to be the generally accepted type of housing. Gradually the formidable requirements of a smoothly operating dome have been met by skillful engineering. Recent innovations have been the use of fluid couplings in the drive, and the development of automatic controls for turning the dome as the telescope moves. It has long been recognized that much of the optical turbulence originates near the ground. Newer telescopes are consequently being mounted as high as 100 feet above the ground. More attention is also being paid to the insulation of telescope buildings and domes, and, in at least one case, temperature controls are being used to minimize external turbulence.

The reflectors I have discussed are of the conventional type and they have one weak point in common. They give good definition only over a relatively small field. This handicap was overcome to some extent by the invention of the Ross correcting lens, first used in the 1930s at Mount Wilson. A dramatic improvement came with the invention of the Schmidt telescope, named for Bernhard Schmidt who devised it about 1930 at the Hamburg Observatory. It is practically free from chromatic aberration and provides fine definition over a wide field. It is greatly to the credit of the amateur telescope makers that they did much of the pioneering work on the Schmidt telescope prior to its professional acceptance. This was done in California as early as 1932 and later in England. New design features were developed and greatly stimulated the adoption of the Schmidt camera for research. The instrument consists optically of a spherical mirror, an aperture stop at the center of curvature, and a weak correcting lens. This type of telescope came into prominent use professionally with the 18-inch at Palomar, the 24-inch at Warner and Swasey Observatory in Cleveland and, later, the 48-inch at Palomar. The 48-inch Schmidt covers an area of 36 square degrees with critical definition, compared to a conventional 36-inch reflector of $f/5$ which does not do as well over a half square degree. The Schmidt telescope is unrivaled for survey studies. Other optical systems for producing large fields, notably the Maksutov system, have been tried with some success, but the Schmidt remains supreme for large-field photography.

The telescopes that have been described thus far are all used primarily for stellar observations. Solar telescopes have quite different requirements, since the light-gathering power of the instrument is not as important a factor as it is in stellar telescopes. Since the sun is so bright, it is possible to analyze its light with much more powerful instruments than can be used for stellar work. Such instruments are ordinarily too large to be attached to a moving telescope. George Ellery Hale played a central role in this field, and the results of his work and subsequent innovations are described in other essays in this volume.

A major consideration in the operation of telescopes is the selection of a suitable site. The amount of clear weather, the transparency of the atmosphere, and the darkness of the sky are all important. The essential requirement is good seeing. The first large observatory located at a really good site was the Lick Observatory, and its choice was largely a matter of luck. James Lick had first chosen Lake Tahoe as the site for his great telescope. Discouraged by reports of the severe winters, he reluctantly turned to Mount St. Helena, California, and directed his confidential agent, Thomas Fraser, to look over the land there. Late in 1874 Fraser met with Lick, reporting on the St. Helena site, and also mentioning Mount Hamilton. Fraser later recalled:

I then told him there was a mountain in Santa Clara County in full view of San Jose that might be suitable and that he should send me to look at it before he settled on St. Helena. He said it could not be 4000 feet high or he would have heard of it before that time. I then went to the Government Office to find out the exact height of the mountain and found it to be 4400 feet. I gave Mr. Lick the information and I may say I never during the time I had been acquainted with Mr. Lick can remember of seeing him so well pleased as he was on hearing that there was such an elevated point in his own County and acting upon my suggestion he sent me to look at Mount Hamilton to see if it was possible to get a road to the summit and if there was any ground on the summit where he could locate the observatory, and if so he would commence work at once.[12]

However, before it was built, the astronomer Sherburne W. Burnham, was employed to make a two-month test of the seeing.

Hale made a careful investigation of possible sites for his solar observatory before selecting Mount Wilson. More recently, sites at Kitt Peak and at Cerro Tololo in Chile were chosen only after surveys extending over two years or more. There is an increased understanding of seeing. Given adequate geographical and meteorological data, a good guess can be made without ever visiting the site. There is now little danger of an observatory being placed in an unsuitable site unless political or other nonscientific considerations dictate its location.

The preceding survey of the development of ground-based telescopes over the past eighty years demonstrates how much the field has changed since the beginning of Hale's work. New materials and new techniques have been ably exploited by the design engineers. Automatic guiding and the automated operation of telescopes have introduced a new dimension whose ultimate role we are only beginning to perceive. Several institutions stand out for their contribution to telescope development. For many years the largest load was carried by the Mount Wilson Observatory through its command of generous financial support, its talented staff, and the imaginative leadership of George Ellery Hale. Today this sphere of activity is shared by the Kitt Peak National Observatory and by other institutions, notably the European Southern Observatory. A number of commercial firms are also participating in exploring new materials for making and finishing mirrors and in developing new techniques in building and operating telescopes.

Technically it is now possible to increase the practical defining, or resolving power, of our telescopes. This field has perhaps been neglected as the limitation has largely been set by the turbulence in our atmosphere. Now the limitation has been minimized in some degree by the intelligent selection of observing sites and by temperature controls over areas in the path of the incident light. But until we can make observations outside the atmosphere, the full defining power of a large diffraction-limited telescope cannot be utilized.

We are on the threshold of this era. Small telescopes have been carried into the stratosphere by balloons. Small optical instruments have hitchhiked into space on rockets, and astronomical observatories are presently orbiting the earth. Initial plans are being considered for putting a large telescope into space, perhaps one of 50 inches. Without the barrier of the earth's atmosphere, the full theoretical defining power of the mirror could be reached. The scientific advantages of a successful effort of this kind would be almost incalculable.

Every great technical conquest brings with it surprising fundamental discoveries. Space technology has revealed the solar wind, the Van Allen belts, celestial x-ray sources, and numerous properties of the moon and the nearer planets. Even as Hale, through his development of greater astronomical instruments, opened new vistas in the universe, so the new space techniques may be expected to reveal phenomena of which we now have no conception. With space telescopes we shall have access to the entire electromagnetic spectrum, limited only by the sensitivity of our detectors. We shall be able to photograph with perhaps ten times the linear definition possible at present. Faint extended objects, now obscured by superimposed light of the sky, will be revealed. If space science and technology continue to develop as they have in the recent past, we shall enjoy these benefits without having too long to wait.

Notes

1. Howard Grubb, "On Great Telescopes of the Future," *Dublin Scientific Transactions, 1,* 5, 1877.

2. Deed of Trust of James Lick, dated July 16, 1874. Lick Observatory Archives, University of California at Santa Cruz.

3. Letter from David Gill, Jr., to Captain R. S. Floyd, Sept. 18, 1876. Lick Observatory Archives, University of California at Santa Cruz.

4. T. R. Robinson, "On the Relative Power of Achromatic and Reflecting Telescopes," *Monthly Notices of the Royal Astronomical Society, 36,* 307, 1876.

5. *Publications of the Lick Observatory, 8,* 7, 1908.

6. Grubb, "On Great Telescopes of the Future," p. 3.

7. T. R. Robinson and Thomas Grubb, "Description of the Great Melbourne Telescope," *Philosophical Transactions of the Royal Society of London, 159,* 144, 1869.

8. Grubb, "On Great Telescopes of the Future," p. 7.

9. Robinson and Grubb, "Description of the Great Melbourne Telescope," p. 144.

10. Grubb, "On Great Telescopes of the Future," p. 8.

11. Ibid., 12.

12. Letter from T. E. Fraser to the Board of Trustees of the James Lick Trust, Nov. 25, 1876. Lick Observatory Archives, University of California at Santa Cruz.

Sir John Herschel in 1867.
Burndy Library.

William Lassell's 48-inch reflector at Malta.
Royal Astronomical Society.

Telescope used by Lord Rosse for observing spiral nebulae.
Lick Observatory.

The arrival of the 36-inch glass at Lick.
Lick Observatory.

James Lick.
Lick Observatory

The original Crossley reflector. This early symmetrical-type mounting was unsuccessfully supported by mercury flotation.
Lick Observatory.

Spiral nebula photographed by James Keeler using the Crossley telescope on November 6, 1899.
Lick Observatory.

The mirror of the 200-inch Hale telescope. This was the first large mirror with a ribbed structure. *Corning Glass Works.*

The 60-inch mirror on the polishing machine, tipped forward for testing. *The Hale Observatories.*

The 98-inch Isaac Newton telescope at Herstmonceux has a solid Pyrex disc.
Royal Greenwich Observatory.

Asymmetric telescopes

The remounted Crossley. This asymmetrical-type mounting built about 1904 is still in effective use. *Lick Observatory*.

The 60-inch torque tube asymmetric telescope on Cerro Tololo, Chile. The large overhang permits convenient access to the Cassegrain focus.
Kitt Peak National Observatory.

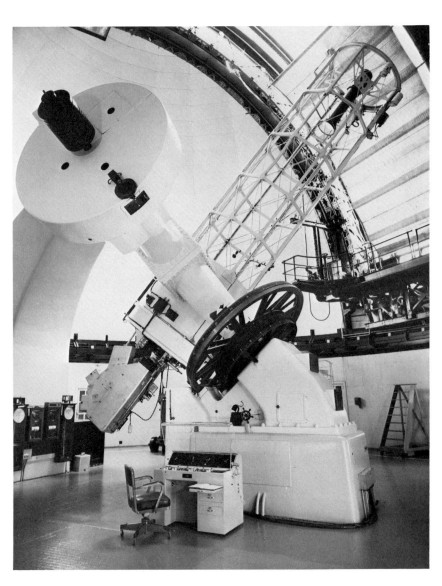

The asymmetric 72-inch reflector of the Dominion Astrophysical Observatory at Victoria.
Dominion Astrophysical Observatory.

The asymmetric 107-inch reflector at the McDonald Observatory, the largest telescope mounted in this fashion.
McDonald Observatory.

Symmetric telescopes

The 60-inch telescope on Mount Wilson, the first important modern telescope with a symmetric mounting. *The Hale Observatories.*

The 100-inch Hooker telescope on Mount Wilson. The symmetric-yoke mounting was adopted for structural reasons.
The Hale Observatories.

The horseshoe-yoke mounting of the 200-inch Hale telescope on Palomar Mountain.
The Hale Observatories.

231   Astronomical Telescopes since 1890

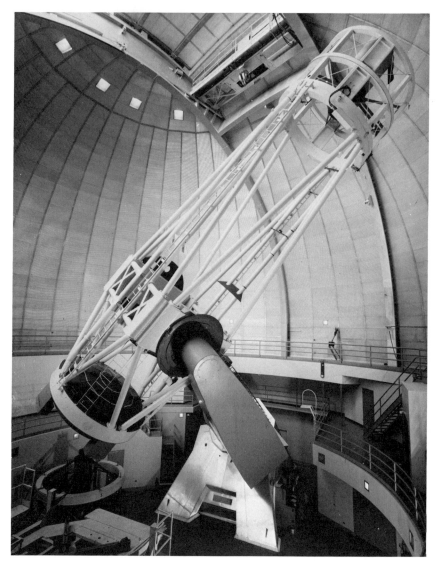

The 120-inch telescope at the Lick Observatory. The symmetric fork-type mounting was made with an extra long fork to permit the attachment of a Cassegrain spectrograph in the optical axis.
Lick Observatory.

C. Donald Shane with the 120-inch Lick mirror.
Lick Observatory.

Observing chairs

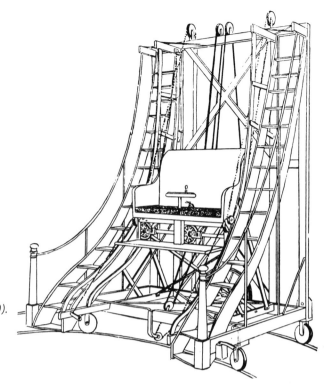

George Bond's observing ladder at the Harvard College Observatory.
Daniel W. Baker, History of the Harvard College Observatory During the Period 1840-1890 (Cambridge: 1890).

The observing chair at the Meudon Observatory, Paris.
Collection of the Museum of Decorative Arts, Paris.

An observer in the cage at the prime focus of the 200-inch Hale telescope. *The Hale Observatories*.

Edwin Hubble observing with the 48-inch Schmidt telescope at Palomar. *The Hale Observatories.*

The balloon ascent of the Stratoscope II, capable of photographing the heavens with three times the detail of the largest earthbound telescopes.
The Perkin-Elmer Corporation.

238 Perspectives

Astronomical Instrumentation in the Twentieth Century

Ira S. Bowen

In 1890, when Hale began his career, a large part of astronomical research was devoted to increasing the precision of measurement of position, including stellar parallax and the separation of double stars. However, Hale's training was primarily in physics and chemistry. At the time rapid advances were being made in spectroscopy and in the techniques for detecting and measuring radiation over a wide range of wavelengths from the ultraviolet to the far infrared. To Hale's alert mind the application of these new techniques to the study of the physical and chemical conditions in stars and nebulae, and to understand the processes occurring in them, was by far the most exciting program for astronomy.

However, much laboratory equipment was designed for use in a fixed position and could not be mounted on a moving telescope. Many instruments were too heavy or big; many could operate only in one orientation. Starting with his Kenwood Observatory and continuing at Yerkes, Mount Wilson, and Palomar, Hale therefore provided machine and optical shop facilities for the construction or adaptation of the instruments for physical measurements at the telescope.

In the early stages of these developments in physics and chemistry, the particular physical data required for comparison with the astronomical results were not available when needed. Hale set up a physics laboratory for these studies as part of the Mount Wilson Observatory. Arthur King's work on furnace spectra and John A. Anderson's studies of exploding wires were both made in this laboratory in an effort to duplicate stellar spectra over a wide range of temperatures.

As astrophysical measurements became an increasing part of the programs of Hale's observatories and of many others, there was a continuing need to extend these observations to fainter and fainter objects. In planning for these advances Hale repeatedly emphasized that the efficiency of the operations often depended as much or more on the proper design of the auxiliary equipment as it did on the aperture of the telescope with which it was to be used. For this reason, Hale included a substantial item for the development of new instrumentation in the original budget for the 200-inch telescope.

Hale early recognized that for many spectroscopic observations of the sun and some of the stars it would be necessary to use a higher dispersion than was obtainable with any spectrograph that could be mounted on a moving telescope. To permit these observations, he built the Snow telescope, completed in 1905. This 24-inch reflector of 60-foot focal length was mounted in a fixed horizontal position while the light was reflected to it with a coelostat. The slit of a Littrow spectrograph of 5-inch aperture and 149-inch focal length was placed at the focus. Besides taking many spectrograms of the sun, this instrument

Walter Adams with the gears of the 100-inch telescope on Mount Wilson. *Photograph by Margaret Bourke-White, LIFE Magazine © Time Inc.*

was used to obtain the first high-dispersion spectrum of a star. In 1905, spectra of Arcturus (α Bootis) and of Betelgeuse (α Orionis) were obtained at a dispersion of 4.3Å/mm, 2½ times that of any previous stellar spectra.[1]

To permit the use of high-dispersion spectrographs and other equipment that could not be mounted on a moving telescope, Hale designed all of his large reflectors with provision for a fixed coudé focus on the polar axis. The soundness of this innovation is shown by the trends of telescope design in the present century. Nine reflectors with apertures over 50 inches were constructed between 1900 and 1945, and only four were provided with a coudé focus. Two of these were of Hale's own design. All of the nine telescopes of this size constructed since 1945 have a coudé focus.

Hale also recognized that if a spectrograph of medium or high dispersion is to have high efficiency, it must have a dispersing system of large aperture. The length of light path in the prism increases proportionally with the aperture; especially at the shorter wavelengths, any glass with high dispersive power absorbs strongly. Consequently, as the aperture is enlarged, the increased absorption of the prism soon nullifies any gains from the larger aperture.

With this basic limitation of prisms in mind, Hale turned to the development of gratings of large size and high efficiency. In 1912 he invited John A. Anderson, who had been in charge of the Rowland engine at Johns Hopkins, to come to Mount Wilson and supervise the design and construction of a large ruling engine. Under the direction of Anderson and later of Harold and Horace Babcock, large gratings of such high efficiency were ruled that by 1955 they had replaced prisms in all of the spectrographs in use at both Mount Wilson and Palomar.

The soundness of this shift to gratings has in recent years been emphasized by the decision of major optical companies, such as Bausch & Lomb and Perkin-Elmer, to invest large sums in techniques and equipment for the ruling of gratings. These companies have also developed methods for making substantial numbers of accurate duplicates of any grating. The availability of many duplicates of each exceptionally fine-ruled grating has made it possible for prisms to be replaced by gratings at most observatories.

Another step in increasing the efficiency of astronomical spectrographs was the development of camera optics of critical definition and extreme speed. This is especially important for the low-dispersion spectrographs used for the faintest objects. One development financed from the instrumentation item in the 200-inch budget was the Rayton lens, designed along the lines of a 4-mm microscope objective to give satisfactory definition through the visual range when used at a focal ratio of $f = 0.84$.[2] Edwin Hubble's concept of the expanding universe was based on Milton Humason's measurements of the radial velocities of very faint and distant galaxies, which were measured on spectrograms chiefly made with this lens.

The invention of the Schmidt camera, first described in a paper by Bernhard Schmidt in 1932, provided for the first time a camera with high speed and critical definition over a much wider range of wavelength than can be attained by any lens.[3] In particular, the thick mirror and aplanatic sphere modifications of the Schmidt camera yield extreme speeds combined with a definition at least as good as that of a fast photographic plate.

These advances made possible the design of such spectrographs as the nebular spectrograph of the 200-inch Hale telescope, whose shortest focus camera operates at a focal ratio of $f = 0.47$.[4] Similarly, the cameras of the coudé spectrograph of this telescope are designed around various modifications of the Schmidt camera. Five cameras of aperture 12 inches and focal lengths ranging from 144 inches to 8.4 inches, that is, focal ratios from 12 to 0.7, are provided. Plans for the 150-inch telescope now under construction at the Kitt Peak National Observatory call for a coudé spectrograph with cameras having an aperture of 24 inches.

Another field of investigation instigated by Hale was the measurement of the infrared radiation of stars. In the 1890s, Ernest Nichols developed one of the first efficient radiometers. In 1898, Hale invited Nichols to bring his radiometer to Yerkes and apply it to this problem. A 24-inch paraboloid of 133-inch focus was set up and the light of the star reflected to it with a plane mirror. Nichols's observations of Arcturus, and Vega (α Lyrae), made in 1898 and 1900, yielded the first definite measurement of the infrared radiation of a star.[5] The sensitivity of radiometers was later improved by Charles Abbot, and he in turn was invited to use the 100-inch on Mount Wilson for stellar measurements. The increased sensitivity, combined with the greater light-collecting power of the 100-inch, enabled him to measure the radiation of a number of stars in a series of nine narrow wavelength bands extending from $\lambda = 0.437$ to $2.224\,\mu$.[6]

The vacuum thermocouple was developed by August Pfund and William Coblentz between 1910 and 1915 and used by them for infrared measurements of stars and planets.[7] In 1920 this technique was taken up by Edison Pettit and Seth Nicholson of the Mount Wilson staff.[8] By 1928 they had built very sensitive Bi-BiSn thermocouples and had used them to measure the total bolometric radiation, including the infrared, of over 100 stars. In 1927 the rate of cooling of the lunar surface during an eclipse was measured with one of these thermocouples.[9] This provided the first evidence for the very low thermal conductivity of the lunar surface and the resultant conclusion that the surface was covered with a dust layer.

The past quarter-century has seen the development of a number of new, very sensitive infrared receivers including the thermistor bolometer, the Golay cell, and a number of photoconductive and photovoltaic cells.[10] These receivers have opened up many fields of astronomical research. They have been used extensively with conventional telescopes, and several major telescopes have been built primarily for infrared observations. Time permits the mention of only a few of these programs.

Richard Shorthill and John Saari, of the Boeing Company, have used a mercury-doped germanium photoconductive cell on the 74-inch Helwan telescope in Egypt to follow the cooling of the lunar surface during an eclipse.[11] In contrast to Pettit and Nicholson's observations, which were able to follow only the cooling of one large area on the moon, the new receiver was so sensitive that Shorthill and Saari were able to cover a raster of 200 scans across the moon with a resolution of 10 seconds of arc in 16 minutes. The data, automatically recorded from a succession of such scans, enabled them to construct a detailed map of the moon showing the thermal conductivity at each point.

In another project, Robert Leighton, Gerry Neugebauer, and Dowell Martz used a lead sulfide photoconductive cell on a 60-inch $f/1$ reflector, constructed for the purpose, to survey the whole sky for infrared objects.[12] In addition to many known stars, a number of objects, so cool that their visual light is undetectable, were discovered. Some of these have a temperature of 1000°K or less, and in a few cases may represent stars in a very early stage of condensation. They were also able to observe and map the infrared radiation coming from the nucleus of our galaxy.

A 28-inch telescope and a 60-inch telescope, located at 8500 feet elevation in the Catalina Mountains near Tucson, Arizona, and designed especially for infrared observations, were used by Harold Johnson and his collaborators for an extensive photometry of a large number of bright stars.[13] Stars of all spectral types were observed at eight or, in some cases, ten wavelength bands extending from the ultraviolet to 10 microns in the far infrared. The effective temperatures of the different spectral classes were then derived from the radiation curves thus obtained.

Finally, Frank Low, using a gallium-doped single crystal of germanium operating at a temperature of 2°K, has measured the radiation at 1 mm wavelength coming from the moon, several planets, and the quasars 3c273 and 3c279.[14]

Another important technical advance of the present century has been the development of precision methods for the measurement of radiation from a star or other source. For this purpose Joel Stebbins, between 1905 and 1930, carried out extensive experiments, first with a selenium cell and then with a photoelectric cell. Again as part of the 200-inch instrument development program, Hale invited Stebbins in 1930 to bring his equipment to the 100-inch and use it for the measurement of stellar and galactic magnitudes. It was soon realized that the magnitudes obtained were of a much higher order of precision than those yielded by the older visual and photographic methods that were found to have errors of up to a factor of 2 at the faintest magnitudes. Indeed, the magnitudes of selected area stars determined by Stebbins and Albert Whitford in the next two decades at Mount Wilson and Palomar were the fundamental standards of magnitude for the color systems they used.[15]

The introduction of the photomultiplier tube, notably the 1P21 at the end of World War II, extended the range of photoelectric measurements to much fainter magnitudes by eliminating most of the noise in the earlier amplifiers. Techniques were then developed for counting the individual electrons ejected from the cathode. This brought the probable error in these observations down to the statistical uncertainty present in counting any large number of random events. With these techniques, it is now possible to measure with some precision the magnitude of a star too faint to be either seen or photographed with the telescope used. The problem is how to find the star!

Originally these photoelectric techniques were used to measure the magnitudes of objects in two or three broad wavelength bands as fixed by available filters in the visual, blue, and ultraviolet ranges. Recently many programs have required a more detailed measurement of a large number of bands of narrower wavelength range. At first these measurements were made with a spectrum scanner. This was a simple spectrometer in which one wavelength band after another could be projected on to a photomultiplier tube by rotating the grating. While valuable results were obtained, this "one band at a time" measurement is very wasteful of the time of a large telescope. The most recent procedure is to use a fixed grating to project a spectrum and to locate along the spectrum a number of photomultiplier tubes, each one receiving its own sharply defined range of wavelengths. Instruments now going into operation have provision for as many as thirty-three photomultiplier tubes that permit the simultaneous counting of the electrons ejected by this number of wavelength bands.

In the course of the use of the photoelectric cell for intensity measurements, it was found that the basic efficiency of the photocathode was some ten to twenty times as great as that of the fastest photographic plate. Thus 10 to 20 percent of the photons striking a cathode eject an electron, while about 100 photons must strike a photographic plate to render one grain developable. In order to take advantage of this greater efficiency of the photoelectric effect in recording faint sources, a long development was initiated by Lallemand in France, followed by McGee in England and Hiltner, Kron, Walker, and the Carnegie Committee in this country.[16]

The basic procedure is to focus the spectrum or star field on a cathode of substantial area. The electrons ejected are then speeded up by an electric potential of several thousands of volts. At the same time they are focused by either an electric or magnetic field with the aid of techniques similar to those used in electron microscopy. At the focus of the electrons either a photographic plate or a phosphor is placed. In the latter case, the image on the phosphor is reimaged on the photographic plate with an optical system. In some cases the intensification of the image is still further increased by repeating the process one or more times.

While the intensification produced by these image tubes is very large, a substantial revision of spectroscopic optics has been required to take full advantage of the gains. Thus, because of the size of the image tube and its accompanying magnets and shields, the tube cannot be placed on the axis of a fast Schmidt camera. It has therefore been necessary in many cases to develop new high-speed optics both for the spectrograph camera and, in the case of tubes forming an image on a phosphor, for the system to reimage this on the plate.[17] However, with equipment now in use it has been possible to reduce the exposure time required to obtain the spectrum of a faint object by a factor of between 10 and 20.

Still another instrument that Hale introduced to astronomy was the interferometer. In Volume 1 of the *Astrophysical Journal* published in 1895, there is a short article written by Hale entitled "On a Photographic Method of Determining the Visibility of Interference Fringes in Spectroscopic Measures."[18] In this article Hale records the observation with a Michelson interferometer of the fringes in the Hα line in a solar prominence and an unsuccessful attempt to observe fringes in the light from the Orion Nebula.

The Fabry-Perot interferometer was developed at the turn of the century. Charles Fabry and Henri Buisson applied this interferometer to the study of emission lines in Orion in 1911 and 1914 and developed techniques for the study of absorption lines by combining the interferometer with a grating in 1905 and 1910.[19,20] These techniques were introduced in the early 1920s at the Mount Wilson Observatory by Harold Babcock, who used them to make the first precise measurement of the wavelength of the green auroral line and to map the infrared solar spectrum.[21]

Within the last decade or two it has been recognized that for very high resolution observations the interferometer is more efficient than a grating. Thus the interferometer can accept light from a much larger area of the sky than is possible for a slit spectrograph of high resolution. Since accuracy increases with the number of quanta recorded, this gives the interferometer a large advantage, especially for studies of objects with extended areas, such as nebulae or galaxies.

Recent technological advances have substantially improved the Fabry-Perot interferometer.[22] The development of multilayer reflecting coats has raised the reflecting power and increased both the efficiency and the resolving power. The development of the photon-counting techniques, already discussed, has greatly increased the accuracy of measurement, especially for very faint sources.

The auxiliary equipment to adapt the interferometer to astronomical observations has also been made much more efficient. G. Courtès has redesigned the optics for making the light from the telescope objective parallel before passage through the interferometer and then for reimaging this light along with the fringes on the plate. With the aid of this equipment, which uses fast Schmidt optics, he has been able to record fringes from H_2 regions whose surface intensity is less than one thousandth of the Orion Nebula.

The interferometer, using the scanning technique, has proved to be valuable for recent studies at high resolution of the shapes of spectrum lines. The scanning over the spectrum line is accomplished by varying the gas pressure and therefore the index of refraction of the space between the interferometer plates. The number of electrons released at the cathode are then counted as the scan progresses. For emission lines it is in general only necessary to isolate the line under study with a suitable filter. This technique has been used successfully by a number of observers in England and by Arthur H. Vaughan in the United States.

Scanning absorption spectra requires that the interferometer be combined with a high-resolution premonochrometer, usually a large coudé-grating spectrometer which is connected so that it scans the spectrum along with the interferometer. This technique was used first in England. In the United States Vaughan and Guido Munch have made use of it to investigate the detailed shape of the Na and Ca interstellar lines in order to obtain the velocity distribution in the clouds of these gases for comparison with the velocities of the hydrogen clouds in the same region as obtained with radio telescopes.

Hale was also instrumental in the introduction of another interference technique to astronomy, namely, the use of the interference between two separated beams for the measurement of star diameters and the separation of very close double stars. In 1919, Albert Michelson was appointed a Research Associate of the Mount Wilson Observatory for the purpose of applying his interference techniques to this problem. The first stellar diameter, of α Orionis, was measured by Francis Pease on December 13, 1920, using the 100-inch telescope with a 20-foot girder to extend the separation between the two interfering beams up to this distance.[23] Further measurements with this equipment yielded values for the diameters of seven stars ranging from 20 to 47 thousandths of a second of arc.

A later attempt to extend the measurements to stars of smaller angular diameter using an interferometer with a beam spread up to 50 feet was only partially successful, although preliminary results suggested values for the diameters of seven additional stars ranging from 8.4 to 34 thousandths of a second of arc. Apparently the difficulties of maintaining the equality of the path length of the two beams to an accuracy of a wavelength or two was too great for the technology of the time.

Further progress in the direct measurement of stellar diameters had to await the development of the intensity interferometer by Hanbury Brown in the 1960s.[24] This instrument obviates the necessity of maintaining great precision in the equality of the two light path lengths and therefore makes it practical to use a much larger separation between the two interfering beams. Brown has recently reported the diameters of fifteen stars brighter than the second magnitude and ranging in stellar type from B_0 to F_s. Diameters between 0.7 and 6.5 thousandths of a second of arc were measured. These measurements, when combined with Pease's diameters of late-type stars of classes K_1 to M_6, provide a valuable check on the surface temperature of stars over a wide range of stellar types.

One instrumental technique that has come to the aid of astronomical research since Hale's day is the high-speed digital computer. It has found use in the reduction of a wide range of astronomical data including, for example, the reduction of position measurements to wavelengths in spectroscopy and the correction for atmospheric absorption in photometric magnitude measurements. Its applications to many theoretical studies are too numerous to mention.

Small- or medium-size computers are now extensively used at the telescope. For example, the multichannel spectrophotometer already described involves the alternate observation of object and sky background for short intervals of time and the subtraction of the two readings to obtain the net radiation of the object. To record manually the reading of several dozen channels even once every few seconds is obviously out of the question. On the other hand, the memory of even a small computer can record and store the data very effectively. The computer can also be used to set the telescope on a prerecorded group of objects and to feed into the telescope controls corrections for atmospheric refraction, telescope flexure, and so forth, thereby greatly speeding up the operation of the telescope, especially when many short observations are scheduled.

Finally, mention should be made of Hale's greatest personal contribution to instrument design, the spectroheliograph that he invented while an undergraduate at M.I.T. This instrument permits one to photograph the sun in monochromatic light and makes possible the study of the gas clouds surrounding the sun. Daily observations with this instrument at a large number of solar observatories have provided much of our knowledge of the solar atmosphere and its changes and motions.

In recent years the spectroheliograph has been modified to measure the radial velocity of the gases and to map the general magnetic field over the sun's surface. Many of these later developments are detailed in Robert Howard's paper in this volume.

In this discussion I have omitted mention of instrumentation for radio astronomy and for observations from above the atmosphere, as these are developments that have occurred almost entirely since Hale's day and are not closely related to the techniques that he initiated or promoted.

From this brief summary of the development of instrumentation in the present century, two facts stand out. The first of these is a direct result of the shift in emphasis from positional astronomy to the study of the physical conditions. Because of this shift, much of modern observational astronomy involves taking the optical techniques of the physical laboratory to the telescope. However, the extreme faintness of many very distant objects has made it necessary to push the sensitivity of receivers and the speed of optical systems to the farthest possible limits. Indeed, at the present time the most sophisticated receivers and the fastest spectrographs are in general to be found at the observatories rather than in the physical laboratories.

The second fact that stands out is that some fifty to seventy-five years ago Hale recognized the importance of nearly all of the basic techniques that now dominate observational astronomy. Furthermore, he took active steps to see that they were fully exploited at that early date at the observatories he directed.

Notes

1. W. S. Adams, "Sunspot Lines in the Spectrum of Arcturus," *Astrophysical Journal, 24,* 69-77, 1906.

2. W. B. Rayton, "Two High-Speed Camera Objectives for Astronomical Spectrographs," *Astrophysical Journal, 72,* 59-61, 1930; and M. L. Humason, "The Rayton Short-Focus Spectrographic Objective," *Astrophysical Journal, 71,* 351-356, 1930.

3. I. S. Bowen, "Schmidt Cameras," *Stars and Stellar Systems,* Vol. 1, *Telescopes* (University of Chicago Press, 1960), pp. 43-61.

4. I. S. Bowen, "The Spectrographic Equipment of the 200-Inch Hale Telescope," *Astrophysical Journal, 116,* 1-7, 1952.

5. E. F. Nichols, "On the Heat Radiation of Arcturus, Vega, Jupiter, and Saturn," *Astrophysical Journal, 13,* 101-141, 1901.

6. C. G. Abbot, "Radiometer Measurements of Stellar Energy Spectra," *Astrophysical Journal, 60,* 87-107, 1924; "Energy Spectra of Stars," ibid., *69,* 293-311, 1929.

7. H. F. Weaver, "The Development of Astronomical Photometry," *Popular Astronomy, 54,* 211-230, 287-299, 339-351, 389-404, 451-464, 504-526, 1946.

8. E. Pettit and S. B. Nicholson, "The Application of Vacuum Thermocouples to Problems in Astrophysics," *Astrophysical Journal, 56,* 295-317, 1922; "Stellar Radiation Measurements," ibid., *68,* 279-308, 1928.

9. E. Pettit, and S. B. Nicholson, "Lunar Radiation and Temperatures," *Astrophysical Journal, 71,* 102-135, 1930.

10. J. Strong and R. F. Stauffer, "Instrumentation for Infrared Astrophysics," *Stars and Stellar Systems,* Vol. 2, *Astronomical Techniques* (University of Chicago Press, 1960), pp. 256-280.

11. R. W. Shorthill and J. M. Saari, "Thermal Anomalies of the Totally Eclipsed Moon of December 19, 1964," *Nature, 205,* 964-965, 1965; "Nonuniform Cooling of the Eclipsed Moon: A Listing of Thirty Prominent Anomalies," *Science, 150,* 210-212, 1965.

12. G. Neugebauer, D. E. Martz, and R. B. Leighton, "Observations of Extremely Cool Stars," *Astrophysical Journal, 142,* 399-401, 1965; B. T. Ulrich, G. Neugebauer, D. McCammon, R. B. Leighton, E. E. Hughes, and E. Becklin, "Further Observations of Extremely Cool Stars," ibid., *146,* 288-290, 1966; G. Neugebauer and R. B. Leighton, "The Infrared Sky," *Scientific American, 219,* 50-65, August 1968.

13. H. L. Johnson, "Astronomical Measurements in the Infrared," *Annual Review of Astronomy and Astrophysics,* Vol. 4 (Palo Alto, Calif.: Annual Reviews, Inc., 1966), 193-206. H. L. Johnson, R. I. Mitchell, B. Iriarte, and W. Z. Wisniewski, "*UBVRIJKL* Photometry of the Bright Stars," *Communications of the Lunar Planetary Laboratory,* University of Arizona, Tucson, *4,* No. 63, 98-110, 1966.

14. F. J. Low, "Lunar Observations at a Wavelength of 1 Millimeter, *Astrophysical Journal, 142,* 1278-1282, 1965; "Observations of 3C273 and 3C279 at 1 mm," ibid., 1287-1289, 1965.

15. J. Stebbins, A. E. Whitford, and H. Johnson, "Photoelectric Magnitudes and Colors of Stars in Selected Areas 57, 61, and 68," *Astrophysical Journal, 112,* 469-476, 1950.

16. W. K. Ford, "Electronic Image Intensification," *Annual Review of Astronomy and Astrophysics,* Vol. 6 (Palo Alto, Calif.: Annual Reviews, Inc., 1968), pp. 1-12, and references given there.

17. I. S. Bowen, "Astronomical Optics," *Annual Review of Astronomy and Astrophysics,* Vol. 5 (Palo Alto, Calif.: Annual Reviews, Inc., 1967), pp. 45-66.

18. G. E. Hale, "On a Photographic Method of Determining the Visibility of Interference Fringes in Spectroscopic Measurements," *Astrophysical Journal, 1,* 435-438, 1895.

19. C. Fabry, and H. Buisson, "Application of the Interference Method to the Study of Nebulae," *Astrophysical Journal, 33,* 406-409, 1911 and C. Fabry and H. Bourget, "An Application of Interference to the Study of the Orion Nebula," ibid., *40,* 241-258, 1914.

20. C. Fabry, "Sur l'Application au Spectre Solaire des Méthodes de Spectroscopie Interférentielle," *Comptes Rendus, 140,* 1136-1139, 1905; and C. Fabry and H. Buisson, "Interférences Produites par les Raies Noires du Spectre Solaire," *Journal de Physique* (Paris), *9,* 197-205, 1910.

21. H. D. Babcock, "A Study of the Green Auroral Line by the Interference Method," *Astrophysical Journal, 57,* 209-221, 1923; and "A Study of the Infrared Solar Spectrum with the Interferometer," ibid., *65,* 140-162, 1927.

22. A. H. Vaughan, Jr., "Astronomical Fabry-Perot Interference Spectroscopy," *Annual Review of Astronomy and Astrophysics,* Vol. 5 (Palo Alto, Calif.: Annual Reviews, Inc., 1967), pp. 139-166.

23. A. A. Michelson, "On the Application of Interference Methods to Astronomical Measurements," *Astrophysical Journal, 51,* 257-262, 1920; and A. A. Michelson and F. G. Pease, "Measurement of the Diameter of α Orionis with the Interferometer," ibid., *53,* 249-259, 1921.

24. H. Brown, "Measurement of Stellar Diameters," *Annual Review of Astronomy and Astrophysics,* Vol. 6 (Palo Alto, Calif.: Annual Reviews, Inc., 1968), pp. 13-38.

Floor plan of Yerkes Observatory designed by Hale.
Astrophysical Journal, 5, April 1897.

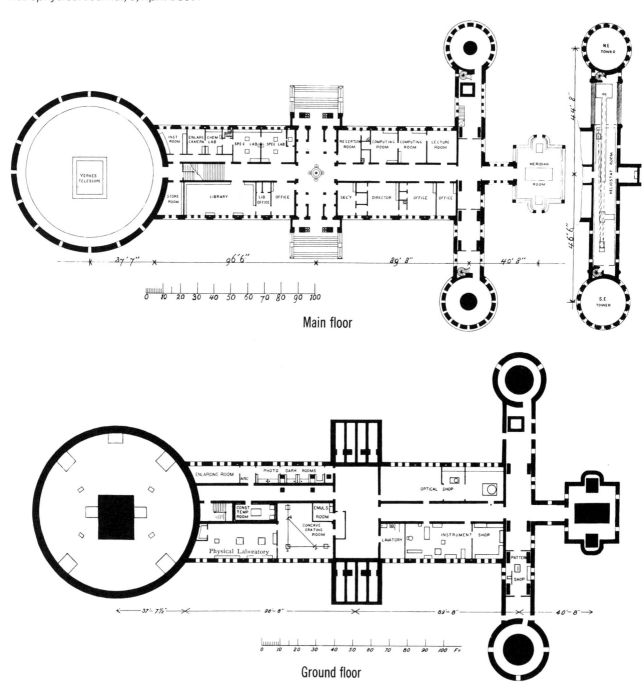

Early Pasadena laboratory of the Mount Wilson Solar Observatory. *The Hale Observatories.*

251 Astronomical Instrumentation in the Twentieth Century

Edwin Hubble on Mount Wilson.
Photograph by Margaret Harwood taken in 1923.

Henry A. Rowland with his ruling engine.
Hale, The Study of Stellar Evolution (Chicago: University of Chicago Press, 1908).

Milton Humason on Mount Wilson.
Photograph by Margaret Harwood taken in 1923.

Humason examining an astronomical plate.
The Hale Observatories.

Bernhard Schmidt in his workshop.
Photograph courtesy of A. A. Wachmann.

Francis Pease using the Michelson interferometer with the 100-inch telescope.
The Hale Observatories.

Robert Millikan and Ira Bowen in the cosmic ray laboratory at the California Institute of Technology.
The Hale Observatories.

255 Astronomical Instrumentation in the Twentieth Century

The Robert R. McMath solar telescope at Kitt Peak.
Kitt Peak National Observatory.

Research on Solar Magnetic Fields from Hale to the Present

Robert Howard

Of all the fields to which George Ellery Hale contributed so much, solar research was his favorite. Often in his letters he wrote of his desire to turn his attention again to the field he loved. It is difficult to estimate the relative influence of Hale on various fields, but certainly his impact on solar astronomy has been very great. In fact, he may be said to be the father of modern solar observational astronomy. For example, the spectroheliograph—which is still one of the most important tools of solar research—was invented by Hale. A relatively recent method of measuring magnetic fields on the surface of the sun by a modification of the spectroheliograph was found to have been suggested by Hale more than fifty years ago. His influence on the nomenclature of sunspot magnetic classifications and chromospheric features still pervades the field.

One of the most interesting and important branches of solar astronomy, and one which Hale originated, is that of solar magnetism. As long ago as 1889, Frank Bigelow suggested that the sun possessed a magnetic field because of the appearance of the corona during eclipse.[1] But Hale was the first person to detect the presence of magnetic fields in any star when he showed that the splitting of spectrum lines in sunspots was due to the Zeeman effect.[2] These were very exciting times in solar astronomy. Judging from the correspondence from that time, one can infer that Hale and his colleagues thought they were on the verge of explaining the sunspot phenomenon. How discouraged they would have been if they could have known how little progress was to be made in understanding the nature and origin of sunspots in the sixty years following the discovery of their magnetic fields.

In 1910 Hale built the 150-foot solar tower telescope on Mount Wilson. With its 75-foot vertical spectrograph it is still one of the leading instruments in solar research. The reason for constructing an instrument with such long focal lengths was simply to get the spectral resolution necessary to measure the general magnetic field of the sun. Incidentally, the 150-foot solar tower is still used for the measurement of solar magnetic fields. Hale recognized at an early stage the need for powerful spectrographs in solar research, and he initiated a project in 1914 to develop large high-quality gratings. The first work with the 150-foot tower used a 6-inch plane grating ruled by Albert Michelson.

In 1912 a concerted effort was begun at the Mount Wilson Observatory to determine whether or not the sun possessed a general magnetic field analogous to that of the earth. The method of measuring the field was straightforward. Spectra of the sun were exposed in the high-dispersion spectrograph. A polarizing analyzer was used, consisting of a Nicol prism and narrow strips cut from a mica quarter-wave plate and oriented compared to the Nicol such that adjacent strips admitted light of opposite circular polarization. Thus a magnetic field oriented parallel with the line of sight would show displacements of a Zeeman-sensitive line in opposite directions from the null position on adjacent strips in the spectrum. The measurement of the plates, then, consisted of measuring the displacements of two adjacent narrow strips of a spectrum line. The dispersion was about 0.2 Å/mm in the third order, and for a sensitive line the displacements for a magnetic field of 10 gauss were only about 0.003 mm. Exposure times were of the order of 30 minutes—which probably washed out most expected velocity-line shifts.

Hundreds of plates were exposed in 1912 and in the years that followed. Great care was taken both in the observations and in the reduction procedure that followed. Often the plates were measured by several people, and, in general, the people who measured the plates were not aware which hemisphere of the sun was represented by the plate they were measuring. The result of thousands of measurements on twenty-six spectrum lines was that the sun possessed a dipole field with a strength of about 20 gauss. The result of 50 gauss, which is often quoted, was observed with only one line. A better average of the lines measured would have been 10 gauss.[3] Later, Robert Langer remeasured many of the original plates with an improved measuring machine and confirmed the first result. The field he measured was 4 gauss.[4] The general field measured in the northern hemisphere at the solar activity minima of both 1912 and 1923 was of negative polarity.

Attempts to measure the general field of the sun continued into the 1930s by Hale's group.[5] A visual technique was employed, which used a tipping plate in front of the eyepiece. In the autumn of 1933, fifty-four sets of such observations gave a magnetic field of $+3.6 \pm 1.7$ gauss. Another set of thirty observations made in the winter of 1948-1949 (close to solar activity maximum) gave a value of -2.0 ± 2.8 gauss. These observations were made at heliographic latitudes of $\pm 45°$.

One of the original reasons for supposing that sunspots possessed magnetic fields was the existence of a spiral iron-filinglike pattern in chromospheric features around some spot groups. In 1941, Robert Richardson showed that such patterns are hydrodynamic rather than magnetic in origin because the direction of the spirals depends not upon the magnetic polarities of the spots but upon the hemisphere in which they are found.[6] This may be an example of the Coriolis force at work.

Harold Babcock employed a Lummer plate with a high-dispersion spectrograph to measure the general field of the sun.[7] Beginning in

1939 and continuing for a decade, he measured magnetic fields at latitudes of ±45°. The results from forty-two sets of plates varied from −6 to −60 gauss. The remaining twenty-four sets showed either no field or small positive values. H. von Klüber used a very similar method to that of Babcock in 1949 and concluded that if there was a field it was smaller than 1 or 2 gauss.[8]

Present methods of measuring magnetic fields on the sun employ photoelectric devices, so it is of interest to trace the development of this technique. Characteristically, the first attempts at photoelectric measures of magnetic fields on the sun were made by Hale in collaboration with Theodore Dunham, John Strong, Joel Stebbins, and Albert Whitford.[9] The state of the art at that time was not sufficiently advanced to make such measures fruitful. It was not until technological advances of the Second World War made more sensitive photometry possible that successful application could be made to the problem of measuring solar magnetic fields. G. Thiessen in 1949 and 1951, using a photoelectric technique, obtained values of -1.5 ± 0.75 gauss and -2.4 ± 0.5 gauss.[10] His observations were made, as was becoming the custom, at latitudes of ±45°, and it is interesting to note that his integration time was 63 seconds, compared to modern values of 0.1 second or less. K. O. Kiepenheuer attempted measurements with an ac photoelectric technique at the 150-foot tower telescope of the Mount Wilson Observatory.[11] He modulated the polarizing analyzer and set his photomultiplier on one wing of a Zeeman-sensitive line. It was necessary to compensate for elliptical polarization, which resulted from reflections within the telescope and therefore varied during the day. His results from observations during 1951 were that there were no polar fields within a limit of 0.6 gauss.

In 1952, H. W. and H. D. Babcock constructed the first "solar magnetograph" at the Hale Solar Laboratory of the Mount Wilson Observatory in Pasadena, and a new era of measurement of magnetic fields had begun.[12] There were two principal advantages of the Babcocks' technique over previous attempts. First, an electro-optic light modulator was used in place of mechanical modulation, thus eliminating sources of false modulation. Second, two exit slits were used—one on either wing of a Zeeman-sensitive line—and the magnetic signal was derived from the ac difference of the outputs of the two photomultipliers. This has the great advantage that, to first order, instrumental polarization does not affect the magnetic signal. There are about fifteen such instruments in operation now all over the world. Numerous refinements have been introduced in the years since its first development, but the principle of the solar magnetograph remains unchanged.

Incidentally, the Babcocks' development of the solar magnetograph was to some extent the outgrowth of work by H. W. Babcock on the measurement of stellar magnetic fields.[13] Using the birefringent property of quartz at the 100-inch telescope on Mount Wilson, Babcock was able to obtain two identical spectra side by side—one in each sense of circular polarization. In this way he made the first observation of a magnetic field in a star other than the sun.

So it was not until fifteen years after Hale's death that a reliable method for measuring magnetic fields on the surface of the sun was developed. As a result of magnetograph observations over the last fifteen years, we have learned a great deal about the formation and evolution of magnetic-field features on the sun and about solar activity. That is not the same as saying that we understand what is going on, but it is nevertheless an advance. What is unfolding for us is a picture that is in some ways beautifully simple and in some ways complex. The magnetic fields are serving to tie together the many apparently diverse phenomena of solar activity. It is unfortunate that George Ellery Hale could not have lived twenty-five years longer to see the real importance of the magnetic fields on the sun, which he spent so much of his effort investigating.

The first obvious question to ask is: What about the general field of the sun? The first observations of the Babcocks indicated the existence of a polar field of the sun with a strength of about 2 gauss and polarity opposite to that of the earth.[14] This field was confined to latitudes above about 55°. The field varied somewhat but remained basically unchanged until the time of the next solar maximum, when it reversed polarity.[15] Since the previous maximum, about 1957, the fields have retained the same polarity as the earth's magnetic field, and it is expected that the solar fields will reverse sometime during the 1968 solar maximum.

The magnetic field at the poles of the sun is not evenly distributed over the solar surface. The 2-gauss value was only an average field. A large fraction or perhaps all the magnetic flux at the poles of the sun is confined to small areas, a few hundred thousand square kilometers or less, where the magnetic field is quite strong, 100 gauss or more. These areas are hotter than the surrounding atmosphere and serve as the "feet" of the giant polar plumes that are seen at solar eclipses to extend hundreds of thousands of kilometers into the solar corona.

We believe now that the basic phenomenon of solar activity is the solar active region, and that the basic ingredient of a solar active region is a bipolar magnetic field. There are usually sunspots associated with an active region. The magnetic fields of an active region are oriented in general so that a line joining the two polarities lies predominantly east-west on the sun. The presence of a magnetic field leads directly to other phenomena. Hydromagnetic waves traveling along the lines of force of the magnetic field heat the chromosphere, causing the bright plages seen in the light of very strong spectrum lines. The arching lines of force above the photosphere in the vicinity of active regions serve to collect material that forms the dark filaments seen on spectroheliograms. Often the magnetic fields in the neighborhood of an active region are not in a simple bipolar configuration but form a complicated arrangement of polarities. When this happens, something gives: magnetic features in the photosphere or lines of force in the corona may rearrange themselves slightly as the result of some instability, and the result is a solar flare.

Active regions appear suddenly, over a period of a day or two, and decay slowly over many weeks. It is assumed that the appearance of an active region represents the rising from below of a bundle of lines of force. There is some observational evidence that this is so.[16] Practically the entire surface of the sun is covered with a network of "supergranulation." These features, identifiable in spectroheliograms or Dopplergrams, have characteristic dimensions of 25,000 kilometers and are superposed on the well-known network of granulation seen in white light with dimensions of the order of 1000 kilometers. This supergranulation network was discovered by Robert Leighton, R. Noyes, and G. Simon at Mount Wilson.[17] The 1 to 2 km/sec motion of the material in supergranules, which is directed horizontally toward the periphery of the cells, is believed to sweep magnetic lines of force to the narrow intercellular space, thus concentrating the field into relatively small areas. Leighton has suggested that the supergranular motions tend to erode the strong magnetic fields of active regions and are, in fact, responsible for their eventual dissolution.[18]

Recently much attention has been directed to the observation of magnetic fields on the sun as seen in high resolution. Most or all of the magnetic flux both inside and outside active regions exists in small bundles where the field is very strong. These bundles are less than a thousand kilometers in diameter and contain fields of many hundreds of gauss. Hale was the first to mention the existence of such features, which he called "invisible sunspots."[19] Recent work has been directed toward the evolution of these features, their number and relative importance, and the effect they may have on the measurements made by a magnetograph.

Magnetograph observations enable us to follow the magnetic fields of an active region long after the chromospheric features have faded from view. In this way we can get a much better picture of the large-scale distribution of magnetic fields than was previously possible. As active regions expand and age, their magnetic fields, measured with the relatively crude angular resolution of the magnetograph, weaken.

They begin to break apart, as seen in a magnetogram. If the level of general solar activity is high, as it was in 1959-1961, these weak fields of old active regions will merge to form an easily observable "background-field pattern." This pattern may cover nearly all the surface of the sun equatorward of 40° latitude, with fields which measure 1 gauss or more. In general, the pattern consists of alternating stripes of different polarity which often cross the equator.

The basic features of this pattern at latitudes below about 20° are very long-lived, showing lifetimes of up to several years. Part of this may be due to the tendency for a feature of one polarity to perpetuate itself because an opposite polarity that appears in the same location will cancel some of the original fields, but a feature of like polarity will reinforce the original feature. But part of the long life of the feature seems to result from a tendency for active regions to reappear at roughly the same solar longitude. The pattern below 20° rotates with a period of 27 days and appears to be associated with the 27-day recurrent geomagnetic storms. The pattern of background magnetic fields at latitudes higher than 20° is twisted to some extent by the differential rotation of the sun. At times large "unipolar" magnetic features form from the joined segments of old active-region fields. These are stretched by differential rotation to form large teardrop-shaped features that migrate slowly toward the solar poles.

In periods of low solar activity, magnetic fields of active regions disappear from view of the magnetograph more readily than during periods of high activity. The background-magnetic fields are an order of magnitude weaker around activity minimum, and it is often very difficult to follow them. Nevertheless, the 27-day rotation period of the pattern of older fields, and the tendency for active regions to reappear at the same longitudes are both well marked.

Satellite observations of the interplanetary magnetic field correlate well with the low-latitude solar magnetic fields below. The interplanetary field may be divided into longitude sectors wherein the field is almost always of the same polarity.[20] The polarity of these sectors correlates well with the polarity of background-magnetic-field features on the sun when one makes a $4\frac{1}{2}$-day correction for the time it takes the solar wind to reach the vicinity of the earth. Thus the background-field pattern on the sun is reflected in the interplanetary magnetic fields 93 million miles away.

The foregoing description is a necessarily brief account of some of the observations of solar magnetic fields made within the last fifteen years. The discovery that magnetic fields are closely linked to the various manifestations of solar activity has proved a means of associating the various phenomena which have in the past seemed largely to be physically unrelated. The simple models that we often use to describe a causal connection between the field and the phenomena are probably only roughly true, if at all; but the complexity of the physical situation has so far precluded a formal mathematical solution of the great majority of the problems that we face in a true understanding of the phenomena. That is another way of saying that, although the magnetic-field observations have added greatly to our knowledge of the phenomena of solar activity, there is still very little that we really understand physically about the many transient features on the sun.

The eleven-year activity cycle itself poses one of the most interesting problems in solar astronomy. Several models have been suggested to explain it, but the most successful of these is that introduced by H. W. Babcock.[21] Babcock suggested that the sun is a magnetic oscillator. If we start with a uniform subsurface meridional magnetic field that emerges from the surface in the polar latitudes with one magnetic polarity exclusively at one pole, then differential rotation will stretch these subsurface lines of force in longitude until the component in the longitude direction is quite strong. If we imagine these fields to be grouped in subsurface ropes, these flux ropes will become twisted by the effects of differential rotation with depth in the atmosphere, and this twisting may provide the instability necessary to bend the lines of force so that buoyant forces may bring them to the surface of the sun. In this way it is possible to explain the usual orientation of active-region magnetic polarities: in general, the polarity that is preceding in the sense of rotation is at a slightly lower latitude than the polarity that follows it.

The first flux ropes to form active regions are those at high latitudes, and as the cycle progresses, spots are formed at progressively lower latitudes, a phenomenon that is indeed observed. Because of the slant of the axis of active regions, the following magnetic polarity drifts preferentially toward the pole as the region ages. Some of the preceding magnetic polarity features drift toward the solar equator and merge with similar features of opposite polarity from the other hemisphere. The general effect of expansion of old magnetic features and differential rotation, then, is to provide a flow of predominantly following-polarity magnetic flux toward the polar regions of the sun. During the course of an activity cycle, the polar fields will build up until the new cycle starts. Then the opposite polarity will begin to drift to the pole until the new flux cancels the old and eventually replaces it. That this reversal of polarity is an effect separate in each hemisphere is indicated by the fact that, at the time of the last activity maximum when the polar fields reversed polarity, these reversals were not simultaneous at the two poles. For some months both poles showed the same magnetic polarity because one pole reversed before the other one did. As the following-polarity features drift toward the poles, they set up the subsurface magnetic fields that provide the sources for the active regions of the next solar cycle. The subsurface fields are naturally reversed in polarity from those of the previous cycle, and so are the polarities of the sunspot groups in successive cycles, as Hale showed many years ago.

Interest in the observation of magnetic fields on the surface of the sun has been increasing steadily in the years since the development of the magnetograph by Babcock. Recent interest has been especially great because of the danger that particles thrown off by solar flares pose for space travelers, or even for passengers in supersonic airliners. It is clear to us now that magnetic fields hold the key to the phenomena called solar activity, and it is a tribute to the genius of George Ellery Hale that he recognized at such an early stage the great importance of these elusive magnetic fields.

Notes

1. F. H. Bigelow, "The Solar Corona, Discussed by Spherical Harmonics," *Contributions to Knowledge,* No. 691, 22 pp., Smithsonian Institution, Washington, D.C., 1889.
2. G. E. Hale, "On the Probable Existence of a Magnetic Field in Sun-spots," *Astrophysical Journal, 28,* 315-343, 1908.
3. H. W. Babcock and T. G. Cowling, "General Magnetic Fields in the Sun and Stars," *Monthly Notices of the Royal Astronomical Society, 113,* 357-381, 1953.
4. R. M. Langer, "Current Investigations of the General Magnetic Field of the Sun," *Publications of the Astronomical Society of the Pacific, 48,* 208-209, 1936.
5. S. B. Nicholson, F. Ellerman, and J. O. Hickox. See the "Annual Report of the Director of the Mount Wilson Observatory," *Carnegie Institution of Washington Year Book No. 33* (Washington, D.C.: 1934), p. 138.
6. R. S. Richardson, "The Nature of Solar Hydrogen Vortices," *Astrophysical Journal, 93,* 24-28, 1941.
7. H. D. Babcock, "A Study of the Sun's Magnetic Field," *Publications of the Astronomical Society of the Pacific, 60,* 244-245, 1948.
8. H. von Klüber, "An Attempt to Detect a General Magnetic Field of the Sun by a Spectrographic Method, Using a Lummer Plate," *Monthly Notices of the Royal Astronomical Society, 111,* 2-17, 1951.
9. G. E. Hale, T. Dunham, Jr., J. D. Strong, J. Stebbins, and A. E. Whitford. See the "Annual Report of the Director of the Mount Wilson Observatory," *Carnegie Institution of Washington Year Book No. 32* (Washington, D.C.: 1933), pp. 143-144.
10. G. Thiessen, "Lichtelektrische Messungen des solaren Magnetfeldes," *Zeitschrift für Astrophysik, 30,* 185-199, 1952.
11. K. O. Kiepenheuer, "Photoelectric Measurements of Solar Magnetic Fields," *Proceedings of the Eleventh Volta Meeting* (Rome: Accademia Nazionale dei Lincei, 1953), pp. 264-266.
12. H. W. Babcock and H. D. Babcock, "The Sun's Magnetic Field, 1952-1954," *Astrophysical Journal, 121,* 349-366, 1955.
13. H. W. Babcock, "Zeeman Effect in Stellar Spectra," *Astrophysical Journal, 105,* 105-119, 1947.
14. H. W. Babcock and H. D. Babcock, "The Sun's Magnetic Field, 1952-1954," *Astrophysical Journal, 121,* 349-366, 1955.

15. H. D. Babcock, "The Sun's Polar Magnetic Field," *Astrophysical Journal, 130,* 364-365, 1959.

16. M. K. V. Bappu, V. M. Grigorjev, and V. E. Stepanov, "On the Development of Magnetic Fields in Active Regions," *Solar Physics, 4,* 409-421, 1968.

17. R. B. Leighton, R. Noyes, and G. Simon, "Velocity Fields in the Solar Atmosphere. I. Preliminary Report," *Astrophysical Journal, 135,* 474-499, 1962.

18. R. B. Leighton, "Transport of Magnetic Fields on the Sun," *Astrophysical Journal, 140,* 1547-1562, 1964.

19. G. E. Hale, "Invisible Sunspots," *Monthly Notices of the Royal Astronomical Society, 82,* 168-169, 1922.

20. J. M. Wilcox and N. F. Ness, "Quasi-Stationary Corotating Structure in the Interplanetary Medium," *Journal of Geophysical Research, 70,* 5793-5808, 1965.

21. H. W. Babcock, "The Topology of the Sun's Magnetic Field and the 22-Year Cycle," *Astrophysical Journal, 133,* 572-587, 1961.

Ferdinand Ellerman, Hale's assistant since Kenwood days, in the solar tower in the 1920s.
Photograph by Margaret Harwood.

Horace Babcock with his father, Harold Babcock, in the Hale Solar Laboratory.
Pasadena Star-News photo, December 26, 1952. Niels Bohr Library.

The 150-foot tower telescope on Mount Wilson.
The Hale Observatories.

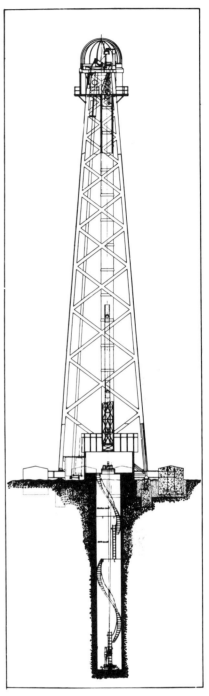

Cutaway view of the 150-foot solar tower.
The Hale Observatories.

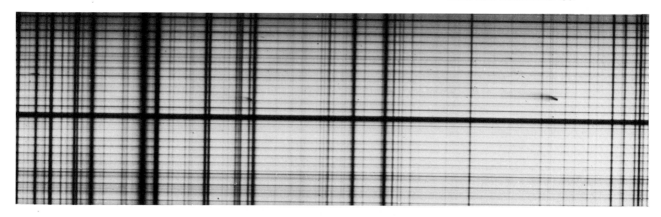

One of the early spectrum plates showing displacements of a Zeeman-sensitive line. The plate is dated December 18, 1913.
The Hale Observatories.

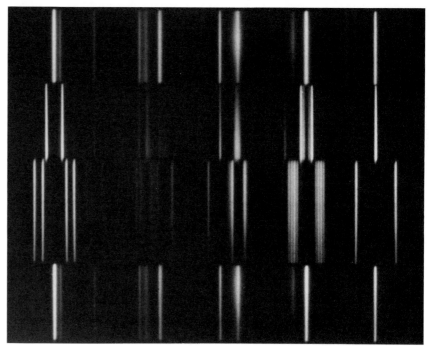

Splitting of spectrum lines by a magnetic field.
The upper and lower strips show lines in the spectrum of chromium, observed without a magnetic field. When the source is subjected to a field, these single lines are split into several components. Thus the first line on the right is resolved by the field into three components, one of which (plane polarized) appears in the second strip, while the other two, which are polarized in a plane at right angles to that of the middle component, are shown on the third strip. The next line is split by the magnetic field into twelve components, four of which appear in the second strip and eight in the third. The magnetic field in sunspots affects these lines in precisely the same way.
Photograph by H. D. Babcock. The Hale Observatories.

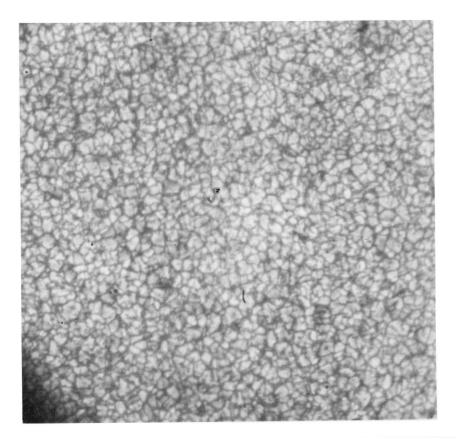

Solar granulation photograph taken by Stratoscope I.
Photograph courtesy of Martin Schwarzschild. Princeton University Observatory.

Hydrogen flocculus falling into a sunspot, June 3, 1908.
The Hale Observatories.

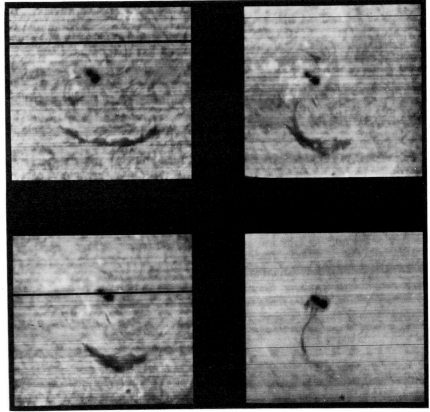

SOLAR MAGNETOGRAMS

THESE MAGNETIC MAPS OF THE SUN'S DISK SHOW THE LOCATION, FIELD INTENSITY, AND POLARITY OF WEAK MAGNETIC FIELDS IN THE PHOTOSPHERE OF THE SUN, APART FROM SUNSPOTS. THE RECORDS ARE MADE AUTOMATICALLY BY A SCANNING SYSTEM THAT EMPLOYS A POLARIZING ANALYZER, A POWERFUL SPECTROGRAPH, AND A SENSITIVE PHOTOELECTRIC DETECTOR FOR MEASURING THE LONGITUDINAL COMPONENT OF THE MAGNETIC FIELD BY MEANS OF THE ZEEMAN EFFECT. A DEFLECTION OF ONE TRACE INTERVAL CORRESPONDS TO A FIELD OF ABOUT ONE GAUSS. THE SMALL DEFLECTIONS OF OPPOSITE MAGNETIC POLARITY NEAR THE NORTH AND SOUTH POLES ARE INDICATIVE OF THE SUN'S "GENERAL MAGNETIC FIELD". THE EXTENDED FIELDS NEAR THE EQUATOR ARISE FROM CHARACTERISTIC "BM" (BIPOLAR MAGNETIC) REGIONS THAT SOMETIMES PRODUCE SPOTS. NORTH IS AT TOP, EAST AT RIGHT.

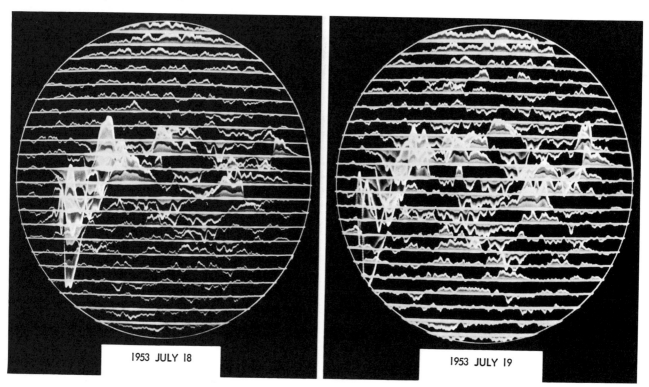

1953 JULY 18   1953 JULY 19

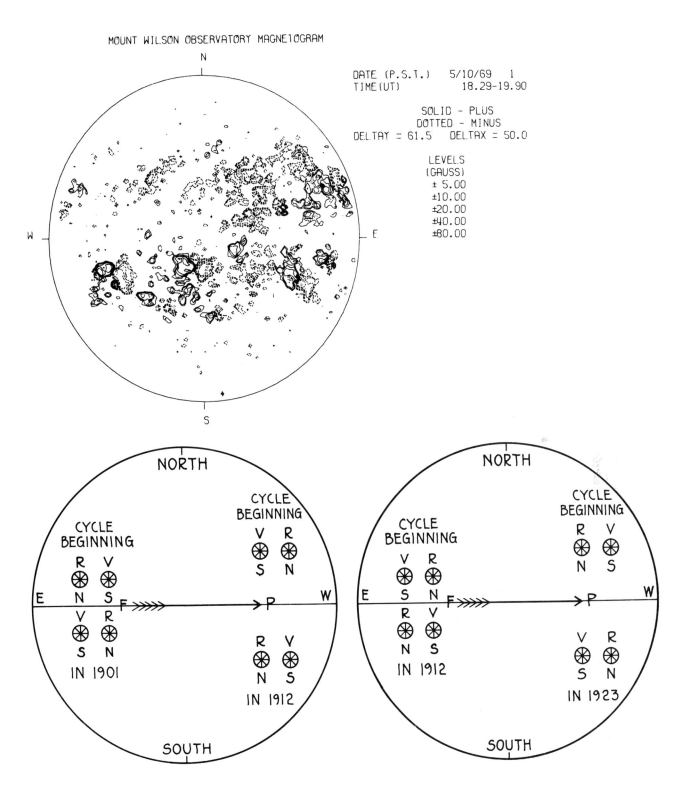

The Hale Observatories.

Founders of the National Academy of Sciences with Abraham Lincoln. Left to right: Benjamin Peirce, Alexander Dallas Bache, Joseph Henry, Louis Agassiz, Abraham Lincoln, Senator Henry Wilson, Admiral Charles Henry Davis, and Benjamin A. Gould.
Painting by Albert Herter in the Academy. National Academy of Sciences Archives.

Hale and the Role of a Central Scientific Institution in the United States

Daniel J. Kevles

Robert A. Millikan once cartooned George Ellery Hale as "the most *restless* flea on the American continent," and so it often seemed, the way Hale kept hatching schemes for new scientific institutions.[1]

"Make no small plans," Hale liked to say.[2] Along with his enormous professional drive were merged an able talent and a humane taste for science. He had invented the spectroheliograph and been elected to the Royal Astronomical Society by the age of twenty-two. He relished English literature—Keats was a pleasant companion for long nights at the telescope—as much as his early editions of Newton. The drive, the reputation, the cultivation, all joined to make Hale a formidably effective promoter of research. He was the son of a wealthy businessman and was naturally at ease marshaling arguments for science to a wide variety of powerful people. Corporate and foundation executives found him a slim bundle of animated enthusiasm, difficult to refuse. Hale had launched the *Astrophysical Journal*, convinced the traction magnate Charles Tyson Yerkes to finance the Yerkes Observatory, and helped persuade the trustees of the Carnegie Institution of Washington to inaugurate and sustain Mount Wilson.

In the decade after the founding of the mountaintop observatory in 1904, Hale's promotional vision gradually extended beyond astrophysics to the challenge of making the United States first-rate in every field of science. At the time, the research community was swiftly expanding (between 1900 and 1914 membership in the American Association for the Advancement of Science jumped more than fourfold, to over 8000, and the special societies showed the same pace of growth). In the universities, scientists were winning more support for their work. In the federal government, the budgets of all the scientific agencies were climbing to unprecedented heights. In industry, a few companies, recognizing that an investment in science could ultimately pay substantial dividends, were opening their own research laboratories; executives at General Electric and A T & T were even urging large corporate contributions to academic science.

For all its growth and ripening possibilities, Hale considered American science, particularly physical science, underdeveloped. Along with many of his colleagues, he saw an acute need for still more public appreciation and financial support. Much more than most of his peers, he perceived that American research had to be rescued from a debilitating emphasis on the gathering of insignificant facts. It is "well-known...," Hale remarked, "that few broad generalizations or fundamental discoveries..., are to be found in the history of this country. We seem to be too much interested in the details and the merely technical elements of investigation, and therefore do not see the woods for the trees."[3]

All the while Hale groped for a way of improving matters, and in 1915 he brought the threads of his thinking together in a slim, programmatic book, *National Academies and the Progress of Research*. The volume contained a wide variety of proposals, almost all of which flowed directly from Hale's own outlook and experience. Hale the cultivated gentleman thought that the public had to be better educated to the intellectual adventure of science (he found the current treatments in the daily press "synonymous with rank sensationalism"). Hale the persuader of businessmen pointed to the potential for funds in the growing interest of manufacturers in research. If scientists were to stress how Michael Faraday's work had laid the foundation for electrical engineering, they could "multiply the friends of pure science and receive new and larger endowments."[4]

Hale the astrophysicist counseled the strategy of his own discipline as a way of ridding American research of its fact-gathering tendencies. He interpreted the prevailing sterility as the product of overly narrow specialization. Few astrophysicists could be counted guilty of this fault, since students of stellar evolution habitually drew upon the findings of physics, chemistry, and mathematics. Hale acknowledged the inevitability of specialization. But interdisciplinary awareness, he argued, would encourage even the most concentrated specialist to confront the "large relationships."[5]

Bolstering the plea for intellectual ecumenicism, Hale the director of the Mount Wilson Observatory called for the encouragement of cooperative research under an institutional framework. Here he distinctly reflected the policy that had always guided the main enterprises of the Carnegie Institution. Shortly after its founding, the trustees announced that the Institution would, wherever appropriate, aim to substitute "organized for unorganized effort," because, as they explained, the "most effective way to discover and develop the exceptional man is to put promising men . . . [to] work under proper guidance and supervision."[6] At Mount Wilson, the kind of cooperation that Hale had in mind was quite evident. There, the staff naturally organized around the telescope, and under Hale's administration the facility had quickly become one of the most successful of the Carnegie Institution's projects. Hale might have cited the vitality of the observatory when he argued that cooperative undertakings stood a better chance of financial support than isolated appeals from individuals; and more, that they encouraged quality research by concentrating many minds upon the various aspects of a few significant subjects. In short, Hale proposed cooperation under institutional auspices precisely because he regarded it as a powerful way of getting American scientists to focus on the large relationships.

It was a broad, multifaceted program, and Hale saw as its most appropriate executor the National Academy of Sciences. He had studied the history of the Academy's European sisters and, impressed by their potent influence, he considered the possibilities of the NAS enormous. In Hale's sure opinion, the Academy had the prestige necessary to persuade both the general and business public of the cultural and practical benefits of research. It also incorporated a disciplinary range broad enough to mitigate the detrimental effects of specialization. Above all, since it represented the whole of American science, the Academy had a rich opportunity to establish standards of high quality in research by promoting cooperation.

To no small extent, Hale was proposing that the Academy do something which it had originally been designed to do. While its charter stipulated that it was to provide advice upon the request of any department of the government, it had been deliberately modeled after its exclusive predecessors in Europe. As elitist organizations of the best scientists in their respective countries, foreign academies managed to stimulate quality science through their memoirs and meetings, and many of the National Academy's founders had hoped that it would play a similar role in the American scientific community. After its creation in 1863 Louis Agassiz, as though anticipating Hale, had rejoiced: Now we have "a standard for scientific excellence."[7]

Over the years the Academy had done little save set a standard for ineffectiveness, at base because it was an academy in the United States. As an exclusive group, it had to endure the resentment of many nonmembers. As a self-perpetuating elite, it wielded little influence, if only because Americans, including scientists, have a way of ignoring quasi-official elites. Moreover, legally centered in Washington, it had to cope with the scientific community's wariness of centralization, and the nation's scientists were spread over too great a geographical area for the Academy's meetings to become anything more than cliquish get-togethers. While its sisters in Europe may have been appropriate to their purpose, the National Academy suffered from the fact that, as citizens of a sprawling, pluralistic democracy, American scientists were simply not like Europeans.

They were, like other Americans, attracted to power, and the Academy's official connection with the government might have given it the status it needed to play an influential role within the research community. But the tie was no more than nominal. The growth of the government's own scientific bureaus had made the Academy's advice increasingly superfluous. In the half-century after its founding, the government had requested help on only 51 occasions, the last in 1908. Moreover, the majority of the membership agreed with the early president of the Academy who said that it would "lose both influence and dignity by offering its advice unasked."[8] In line with that thinking, the Academy, a private organization, refused to ask the government for money, fearing that with appropriations would come political control. Its total income, derived from various bequests, was small. Election to the organization might be a high honor; without money, the Academy could hardly give honor the leverage it needed to shape American science.

Hale knew just how many obstacles the Academy had to overcome before it could act effectively. For the sake of both stimulating the nation's scientists and keeping them abreast of progress outside their own specialties, he had already managed to push the membership into publishing a *Proceedings.* He had also won their agreement that the Academy needed a building—and not only for its own self-respect: a permanent home given over to lectures and exhibits could call the public's attention to the importance of both science and the Academy itself. But Hale had yet to bulldoze the Academy out of its traditional bedrock difficulties. He might aim at the patronage of practical manufacturers; the membership had repeatedly refused to admit engineers or even Thomas Edison himself to their august fellowship. He might cherish visions of the Academy as an action-prone force in American science; many of its members objected to centralization and, in the contemptuous judgment of one, preferred to go on operating as nothing more than a "blue ribbon society." The elect aside, according to Hale's troubled report many nonmembers still considered the elitist Academy "a menace to true democracy."[9]

Hale may have been Hamiltonian enough to stand by the idea of an Academy; he was sufficiently Jeffersonian to realize that the institution could never do much without becoming more representative of all the nation's scientists and demonstrating more concern with their general welfare. He had played a principal role in the decision, made a few years before, to raise the limit on annual elections from five to ten. Now he argued, successfully, for a jump to fifteen. His program also called for the Academy to cooperate with local and national societies, a thrust that would at once mitigate the traditional geographical difficulty and get around the exclusionist objection. Inclusiveness aside, the Academy could surely enlarge its influence by dispensing more money for research, particularly to promising young men.

A building, cooperative activity, aiding research—all would require vastly more funds than the Academy had in its treasury. By the time his book, *National Academies and the Progress of Research,* was published in 1915, Hale had already tried soliciting an endowment from the Carnegie Corporation of New York, but the president of the Corporation had turned down the request. In his blunt explanation, the Academy would deserve no such support until it had ceased demanding deference and had gone much further toward winning the actual respect of the government. The charge merely ratified what Hale already knew from his study of science in Europe. As he noted in his book, "to accomplish great results," any academy had to "enjoy the active cooperation of the leaders of the state."[10]

All Hale's ambitions might rest on winning this happy status; he shared the membership's traditional wariness of thrusting the government's official scientific adviser upon the government. To Hale's mind, the Academy could, upon request, recommend appointees for federal scientific posts and advise on responsibilities that cut across those of single federal agencies. Hale was unwilling to press for more major responsibilities until the Academy's "standing and prestige" were more fully developed.[11] Otherwise, it would become a mere political supplicant.

But Hale's book had hardly appeared before preparedness for war had given the Academy an opportunity to seek the necessary standing and prestige. In a national emergency the NAS could approach the government without fear of political taint, and in April 1916, at Hale's own urging, it did just that. In response, President Woodrow Wilson asked the Academy to organize an arsenal of science for the nation's defense. The request set the organization on the way to winning that item crucial to Hale's program: the active cooperation of the leaders of the nation.

The Academy quickly formed and went on to develop a National Research Council. The NRC's objective was to encourage both pure and applied research for the ultimate end of the national security and welfare. Its strategy for the task was to promote cooperation among all the research institutions of the country. Its membership was to be composed of leading scientists and engineers from the universities, industry, and the government, including the military. Broad cooperation, industrial contacts, federal representation — the NRC incorporated the principal points in Hale's book, and Hale himself could jubilantly say of its creation: "I really believe this is the greatest chance we ever had to advance research in America."[12]

After the diplomatic break with Germany, the Council concentrated less on advancing research than on solving military problems, and by early 1918 it had demonstrated how effective a federation of research interests could be. At least to an extent, the NRC had persuaded a number of traditionally rival federal agencies to cooperate. It had helped funnel military work not only into industrial but university laboratories. And, no small accomplishment for an organization dominated by professors, it had also won the respect of many bureau chieftains in the Army and Navy. Hale, who through it all never forgot his ultimate promotional goal, was certain that the Council's "close contact" with governmental officials as well as with influential industrialists and engineers, had created "many new friends for pure science."[13] The Rockefeller Foundation had asked the NRC to develop a plan for the endowment of research in physics and chemistry; the Carnegie Corporation, impressed at last, had expressed a serious interest in funding the construction of a home for the Academy. The Council's "coalition of widely different interests...," Hale concluded "is certain to be beneficial to research in all of its aspects."[14]

The peacetime prospects of the NRC were made all the more promising when in May 1918 the President issued an executive order—Hale wrote the initial draft—requesting the Academy to perpetuate the Research Council. Without the order, the Council would have had no guarantee of governmental cooperation after the war. With it, future presidents would be bound to appoint representatives from the federal establishment. Aside from this pledge of ongoing participation by the leaders of the state, the document outlined a variety of peacetime functions for the NRC which, together, added up to the stimulation of pure and applied research, particularly through the encouragement of cooperative efforts. When the signed order arrived from the White House, Hale could rejoice: "We now have precisely the connection with the government that we need and... it will be our own fault if we do not make good use of it."[15]

Making good use of it meant going ahead on the basis of the ideas outlined in *National Academies and the Progress of Research*: specifically, funding the Council and setting it up so as to encourage cooperation. Hale argued the economic importance of science to appeal to corporate and foundation America. In the spring of 1918 the NRC created an Industrial Advisory Committee of "strong men," in the proud words of one report, "with the imagination to foresee the general benefits" of pure research.[16] Drawn from the boardrooms of the nation's major industries, the members were to lend their prestige to what Hale frankly called a propaganda program for science. Within a year of the executive order, the NRC had its principal financial gifts in hand. The Carnegie Corporation, of which Elihu Root was a powerful trustee, donated $5,000,000, part of it for an endowment, the rest for the construction of the coveted building. From the Rockefeller Foundation came a grant of $500,000 for postdoctoral fellowships in physics and chemistry.

In 1919, while the fund-raising was in progress, the Council was organized to promote cooperation. Its Division of General Relations included slots for the federal representatives appointed by the President under the terms of the executive order. Its divisions of science and technology brought together practitioners from a wide variety of disciplines, both as individuals and as representatives of the country's numerous professional societies. In the course of it all, Hale emphasized that the Council would not be administratively dictatorial. Scientists, he knew, were exceedingly jealous of their prerogatives and disliked being told what kind of research to do or how to go about it. Sensitive to this feeling, Hale had even written an assurance into the executive order itself. In all "co-operative undertakings" the Council was "to give encouragement to individual initiative, as fundamentally important to the advancement of science."[17] Well-funded and shrewdly organized, the NRC emerged from the war as precisely the kind of national institution that, to Hale's mind, could go a long way toward promoting scientific research of high quality in the United States.

But despite the grandness of the structure, despite the assurance of individual freedom, measured against the full sweep of its architect's visionary purpose, the peacetime National Research Council was largely a failure. As a body with the principal aim of promoting cooperation, the NRC faced obstacles more numerous than those that had traditionally paralyzed the Academy. Most American scientists were still spread over so great a geographical area as to make active cooperation mechanically quite difficult. More important, the vast majority of scientists were concerned with their own special research interests, and, whatever Hale's belief in the value of interdisciplinary cooperation, many disciplines did not readily lend themselves to cross-fertilization from other fields. In addition, cooperation within a discipline like physics, at least before the advent of particle accelerators, was much less natural than in one like astronomy. During the twenties and thirties, the National Research Council managed to stimulate almost no important cooperative research. Its magnificent building in Washington echoed with the sound of unrealized plans and reminded one critic of a "marble mausoleum."[18]

To overcome the obstacles it faced, the Research Council would have needed the kind of authority which comes only from a federal mandate, and that authority Hale emphatically did not want or try to get. Whatever his desire for federal patronage, he emerged from the war still fearing too close a connection with the government. The government might have representatives on the NRC; nominations to the Council lay in the hands of the Academy, which, as Hale said, made it "possible to eliminate all questions of political preferment." The Council might have what Hale deemed the "advantage of Government cooperation"; it remained, like its parent, a private organization without the force of public authority. So far as Hale was concerned, the NRC would have to put its program across to the scientific community not by administrative direction but by "moral pressure."[19]

Had moral pressure been married to money, the NRC might have been more effective. But like his ancestors in the Academy, Hale regarded federal appropriations for science with distinct coolness, and he was content, as he said, to advance research in the United States by depending on "private funds for a long period in the future."[20] Before World War II the NRC never got enough private money to support many cooperative projects, to further interdisciplinary communication, to help specialists buy research assistance, equipment, and time away from teaching. In short, private funding turned out to be inadequate to the scope of Hale's vision. In a country as large as the United States, probably only the federal government could have afforded to promote science on the grand scale which Hale had conceived.

But whatever the flaws in Hale's overall scheme, judged by his goal of improving the *quality* of American research, he had fastened on a significant institutional course. He saw that science in the United States would benefit from some sort of centralized institutional stimulus. At the same time, he recognized both that any centralized scheme would have to respect local vitality and independence, and that, to preserve individual initiative, it would have to be more permissive than directive. In terms of his ultimate goal, Hale's was a perception of double strength. He knew that, like the countries of Europe, the United States needed some national system by which standards of excellence in research could be both set and enforced. And, no less important, he understood that, whatever the details of the system, it could work only if it were adapted to those national characteristics that made America different from Europe.

The acuity of Hale's vision found confirmation in the National Research Council's postdoctoral fellowship program funded by the Rockefeller Foundation. Of the 19,000 American scientists who took their Ph.D.'s in the interwar years, only 883, fewer than 5 percent, were awarded these fellowships; yet, as both historians and scientists agree, somehow the program was profoundly important to the eventual development of first-rate science in the United States.[21] What made it so effective was the way it operated. With the procedure of selection located in a central organization, the NRC, the fellowship boards were composed of some of the nation's best scientists. The members doubtless judged applicants on the basis of how they measured up against a high standard of quality, and the distinction of the awards themselves spotlighted just what kind of work a good young scientist ought to be doing. Once designated, the fellows won a year or two free of teaching duties, a rare and important luxury for a neophyte physicist, chemist, or biologist, fired with eagerness to mine the frontiers of his discipline full-time. More important, they were free to migrate from their home universities to work under the leaders in their specialties at home and abroad; as a result, they evidently developed a more refined taste for the kind of research that was more significant than mere details and technicalities. The NRC fellows finally settled into teaching posts, and there they seeded more than one generation of students with the same sensitivity for asking the right questions of nature.

The entire process called to mind the policy laid down by the trustees of the Carnegie Institution: The "most effective way to discover and develop the exceptional man is to put promising men ... [to] work under proper guidance and supervision." At once the heir and the executor of that policy, George Ellery Hale was more than a remarkable creator of scientific institutions and a formidable promoter of research. He was also a shrewd and penetrating analyst of what a national organization might, indeed had, to do, if the United States were to take its place in the first rank of the world of science.

Notes

1. Millikan to Greta Millikan, July 19, 1916, Robert A. Millikan manuscripts, Archives of the California Institute of Technology, Pasadena, Calif., Box 53. Parts of this essay appeared previously in somewhat different form in *Isis,* and I gratefully acknowledge the permission of the editors to reprint them here. See "George Ellery Hale, the First World War, and the Advancement of Science in America," *Isis, 59,* 427-437, 1968.

2. Frederick H. Seares, "George Ellery Hale: The Scientist Afield," *Isis, 30,* 244, May 1939.

3. Hale to Simon Newcomb, March 21, 1906, Library of Congress, Washington, D.C., Simon Newcomb manuscripts, Box 25.

4. *National Academies and the Progress of Research* (Lancaster, Pa.: New Era Printing Co., 1915), pp. 115, 130-131.

5. Ibid., pp. 100-101; for a more explicit expression of this view, see Hale to Newcomb, January 5, 1899, March 10, 1906, Simon Newcomb manuscripts, Box 25.

6. "Proceedings of the Executive Committee," *Carnegie Institution of Washington Yearbook, No. 1, 1902* (Washington, D.C.: 1903), p. xxxix.

7. Agassiz to Alexander Dallas Bache, May 23, 1863, in Nathan Reingold, ed., *Science in Nineteenth Century America: A Documentary History* (New York: Hill and Wang, 1964), p. 210.

8. See the list of the Academy's accumulated services to the government in Hale, *National Academies,* p. 53; the early president was Othniel C. Marsh, who is quoted in Charles Schuchert, "Othniel C. Marsh," *National Academy of Sciences Biographical Memoirs, XX,* 30, 1939.

9. Edwin Grant Conklin to Hale, March 28, 1913, George Ellery Hale Papers, the Hale Observatories, Pasadena, Calif., Box 11; Hale, *National Academies,* p. 22.

10. President Henry S. Pritchett to Hale, February 3, 1913, copy in Edwin Grant Conklin manuscripts, Archives of Princeton University, Princeton, N.J., National Academy of Sciences file; Hale, *National Academies,* p. 53.

11. Hale, *National Academies,* pp. 83, 164-165.

12. Quoted in Helen Wright, *Explorer of the Universe: A Biography of George Ellery Hale* (New York: Dutton, 1966), p. 288.

13. Hale to Franz Boas, March 14, 1918, National Research Council manuscripts, National Academy of Sciences-National Research Council, Washington, D.C., Institutions file.

14. Hale to Arthur Schuster, April 18, 1918, Hale Papers, Box 47.

15. Hale to James R. Garfield, May 16, 1918, National Research Council manuscripts, Hale file.

16. Council of National Defense, *Second Annual Report, 1918* (Washington, D.C.: U.S. Government Printing Office, 1918), p. 63.

17. Hale's draft of the order is attached to David Houston to Woodrow Wilson, 30 April 1918, Woodrow Wilson Papers, Library of Congress, Washington, D.C., File VI, Case 206.

18. James McKeen Cattell, "The Organization of Scientific Men," *Scientific Monthly, 14,* 576, June 1922.

19. Hale to James R. Angell, August 13, 1919, Hale Papers, Box 3; Hale to Arthur Schuster, April 18, 1918, Hale Papers, Box 47; Hale to John C. Merriam, February 5, 1918, National Research Council manuscripts, Hale file.

20. Hale to Willis R. Whitney, June 19, 1918, Hale Papers, Box 43.

21. See Myron Rand, "The National Research Council Fellowships," *Scientific Monthly, 73,* 75, August 1951.

Members of the National Research Council staff during World War I, including R. A. Millikan, second from left.
California Institute of Technology Archives.

EXECUTIVE ORDER

The National Research Council was organized in 1916 at the request of the President by the National Academy of Sciences, under its congressional charter, as a measure of national preparedness. The work accomplished by the Council in organizing research and in securing cooperation of military and civilian agencies in the solution of military problems demonstrates its capacity for larger service. The National Academy of Sciences is therefore requested to perpetuate the National Research Council, the duties of which shall be as follows:

    1. In general, to stimulate research in the mathematical, physical and biological sciences, and in the application of these sciences to engineering, agriculture, medicine and other useful arts, with the object of increasing knowledge, of strengthening the national defense, and of contributing in other ways to the public welfare.

    2. To survey the larger possibilities of science, to formulate comprehensive projects of research, and to develop effective means of utilizing the scientific and technical resources of the country for dealing with these projects.

    3. To promote cooperation in research, at home and abroad, in order to secure concentration of effort, minimize duplication, and stimulate progress; but in all cooperative undertakings to give encouragement to individual initiative, as fundamentally important to the advancement of science.

    4. To serve as a means of bringing American and foreign investigators into active cooperation with the scientific and technical services of the War and Navy Departments and with those of the civil branches of the Government.

    5. To direct the attention of scientific and technical investigators to the present importance of military and industrial problems in connection with the war, and to aid in the solution of these problems by organizing specific researches.

    6. To gather and collate scientific and technical information, at home and abroad, in cooperation with governmental and other agencies, and to render such information available to duly accredited persons.

Effective prosecution of the Council's work requires the cordial collaboration of the scientific and technical branches of the

Government, both military and civil. To this end representatives of the Government, upon the nomination of the National Academy of Sciences, will be designated by the President as members of the Council, as heretofore, and the heads of the departments immediately concerned will continue to cooperate in every way that may be required.

Woodrow Wilson

The White House,
11 May, 1918.

Executive order issued by President Woodrow Wilson, May 1918.
The National Archives.

We knew he was a great figure in science, but felt that he could have been equally great at almost anything else. For Nature had not only endowed him with those qualities that make for success in science — a powerful and acute intellect, a reflective mind, imagination, patience and perseverance — but also in ample measure with qualities which make for success in other walks of life — a capacity for forming rapid and accurate judgments of men, of situations, and of plans of action; a habit of looking to the future, and thinking always in terms of improvements and extensions; a driving power which was given no rest until it had brought his plans and schemes to fruition; eagerness, enthusiasm, and above all a sympathetic personality of great charm.

Sir James Jeans

Nebulosity in Monoceros, NGC 2264. Photographed in red light with the 200-inch Hale telescope.
The Hale Observatories.

Index

Numbers in italics refer to pages with illustrations.

Abbot, Charles Greeley, *50, 70, 76*, 241
Abney, William, 137
Academy of Athens (Plato's Academy), 186
Academy of Sciences, Paris, 73, 119, 124, 178, 181, 183, 184
 Comptes Rendus, 123, 183
Adams, Walter Sydney, 43, 44, *50, 71, 76,* 99, *100, 109,* 159, 170, 198, 200, *238*
Agassiz, Louis, *272*, 275
Alexandrian Museum, 186
Allen Academy, 2
American Association for the Advancement of Science, 19, 207, 273
American Astronomical and Astrophysical Society, 21, 39
American Astronomical Society, 21, 22
American Chemical Society, 189
American Telephone and Telegraph Company, 273
Ames, Joseph Sweetman, *76*
Anderson, John A., 45, *239*, 240
Arago, François, 117
Arrhenius, Svante, 179
Associated Press, 185
Astronomy and Astrophysics, 21
Astrophysical Journal, The, 21, 22, *39,* 244, 273

Babcock, Harold D., 45, *76,* 240, 245, 258, 259, *260, 266*
Babcock, Horace W., 45, 240, 259, 260, 263, 264, *266*
Bache, Alexander D., *272*
Backlund, Johan Oskar, *76*
Backus, *50*
Ball, Robert, 17
Banks, Joseph, 178
Barnard, Edward E., 18, *33, 35, 50, 76*
Bausch & Lomb Optical Company, 240
Bavarian Academy of Sciences, Munich, 186
Beacon, The, 14
Belopolsky, Aristarch Apolonovich, *76*
Bennett, N. E., *33*
Berlin, University of, 20
Berlin Academy of Sciences, 190
Berry, Edward R., 201
Bessels, Emil, *176*
Bigelow, Frank, 257
Billings, John Shaw, *176*
Boeing Company, 242
Bohr, Niels, 198
Bond, George Phillips, 234
Bosler, *76*
Boston Public Library, 9
Bowen, Ira S., *109,* 198, 239, *254*
Brackett, Frank Parkhurst, *76*
Brashear, John, 9, 18, 20, *24, 33,* 98, 128, 213
Bridge, Norman, 87
Bridge (Norman) Physical Laboratory, *92,* 198
Brooks, William R., *33*
Brown, Hanbury, 246
Buisson, Henri, 244
Bunsen, Robert, 118
Burbank, J. C. B., 136
Burnham, Sherburne W., *8,* 18, 218

California Institute of Technology, 87, *90, 91, 92, 93,* 99, 107, 207, *254*
Callendar, Hugh Longbourne, *33*
Calver, George, 210, 213
Campbell, William Wallace, 18, *76*
Carnegie, Andrew, 42, 45, *60, 61, 62, 63,* 98, 193, 194
Carnegie, Louise W. (Mrs. Andrew), *62*
Carnegie, Margaret, *62*
Carnegie Committee, 244
Carnegie Corporation of New York, 73, 276, 278
Carnegie Institution of Washington, 42, 43, 44, 99, 193, 194, 201, 273, 274, 281
Carrington, Richard Christopher, 157
Cassegrain, Giovanni D., 199
Cerro Tololo Inter-American Observatory, 215, 218, *227*
Chant, Clarence Augustus, *76*
Chicago, University of, 18, 19, 20, 42, 193
Chicago Manual Training School, 2
Chrétien, Henry, *76*
Clark, Alvan, 18, 19, *29,* 127, 128, 193
Clark (Alvan) & Sons, 210, 213
Clark, George, 127, 128, 131, 132
Clarke, Frank Wigglesworth, *176*
Cobb, Henry Ives, 20
Coblentz, William, *76*, 241
Coffin, John Huntington Crane, *176*
Collins, William Henry, *33*
Columbian Exposition, Chicago, 1893, 28, 72
Condorcet, Antoine, 186
Conklin, Evelina; see Hale, Evelina (Mrs. George Ellery)
Coolidge, Calvin, 85
Copley Medal, 2
Corning Glass Works, *104, 105,* 201, 212
Cortie, Aloysius Laurence, *76*
Cotton, Aimé Auguste, *76,* 163
Courtès, G., 245
Crew, Henry, *33*
Crookes, William, 155, 162
Cross, Charles, 9
Crossley, Edward, 210
Cunningham, Susan, *33*

Dalton, John Call, *176*
Darwin, Charles, 151, 187
Davidson, Anstruther, *61*
Davis, Charles Henry, *272*
Davy, Humphrey, 178, 185
Day, Arthur L., 201
De La Rue, Warren, 118
Deslandres, Henri Alexandre, *76*
Dewar, James, 185
Dominion Astrophysical Observatory, *228*
Doolittle, Charles Leander, *33*
Draper, Henry, 210
DuBridge, Lee Alvin, *109*
Dunham, Theodore, 259
Dunn, Gano, 85
Dupree, Hunter, 46
Dutton, Clarence Edward, *176*
Dyson, Frank, *76*

Eddington, Sir Arthur Stanley, 198
Edison, Thomas Alva, 276
Einstein, Albert, 66, 67, 69, *101,* 105, 194, 197
El Karakat, 194
Ellerman, Ferdinand, 21, *33, 35, 50, 51, 76,* 170, *266*
Emden, Jacob Robert, 170

289   Index

Endlich, *176*
Engelmann, George, *176*
European Southern Observatory, 214, 218
Evershed, John, 170
Eversheim, 76

Fabry, Charles, 76, 244
Faraday, Michael, 155, 162, 178, 185, 186, 190, 191, 274
Fath, Edward Arthur, 76
Faye, Hervé, 157
Ferrel, William, *176*
Fleming, Mrs. Williamina Paton, 76
Flexner, Simon, 97
Flint, Albert Stowell, 33
Floyd, Richard S., 209
Foucault, Léon, 214
Fowle, Frederick Eugene, Jr., 76
Fowler, Albert, 76
Fox, Philip, 76
Fraser, Thomas, 217
Freeman, James E., 85
Frost, Edwin Brant, 19, *33*, 35, 76
Furness, Caroline Ellen, 33

Gale, Henry Gordon, 44, 76
Galilei, Galileo, 44, 155, 194, 198
General Electric Company, 201, 273
Geological Society of Great Britain, 178
Gibbs, Josiah Willard, 179, 191
Gill, David, 209
Goethals, George W., 190
Goodwin, Harry Manley, 2, 9, *10*, 35, *40*, 97, 99
Gould, Benjamin Apthorp, *176, 272*
Grove, William, 118
Grubb, Howard, 209, 210, 211, 213, 214, 216
Grubb, Thomas, 213, 214
Grubb Parsons Company, 215
Guyot, Arnold, *176*

Hagen, Father J. G., 33
Hale, Evelina (Mrs. George Ellery), 18, *27*, *33*, *62*, 78, *109*
Hale, George Ellery,* viii, *3*, *10*, 25, *33*, *34*, 35, *38*, *40*, 47, 50, 51, 52, 60, *61*, 70, 76, 83, 86, 92, *108*, *112*, 204, 207, 209, 211, 217, 218, 219, 239, 240, 241, 243, 244, 246, 247, 250, 257, 258, 259, 260, 261, 264, 273, 274, 275, 276, 277, 278, 279, 280, 281, 286
Hale, George E. (nephew), 97
Hale, Margaret, *38*, *62*
Hale, Martha, 27
Hale, William B., *27*, 127
Hale, William Ellery (father), 2, *27*, 42, 45, 186, 188
Hale, William E. (son), *38*
Hale Centennial Symposium, 207
Hale Observatories, 99, 207; *see also* Mount Wilson and Palomar Observatories
Hale Solar Laboratory, 45, 97, *100, 101,* 259, *266*
Hall, Asaph, 33
Halley, Edmund, 117
Hamburg Observatory, 217
Hamy, Maurice Theodore, 76

*The pages refer only to the text of Part 3 and the photographs throughout the book.

Harper, William Rainey, 19, 20, *32*, 193
Harper's Magazine, 98
Hartmann, Johannes Franz, 76
Harvard College Observatory, 9, *13*, 21, 77, 127, 131, 137, 138, *140*, *141*, *142*, 234
Harvard Dry Plate Company, 137
Haussmann, 76
Hedrick, Father John, *33*
Henry, Joseph, *176*, 272
Henry, Mary (daughter of Joseph), *176*
Hepperger, Josef, 76
Herschel, Sir John, 156, 157, 213, *220*
Herschel, Sir William, 155, 156, 210, 213
Hertz, Heinrich, 191
Hilgard, Julius Erasmus, *176*
Hills, 76
Hiltner, William Albert, 244
Holden, Edward S., 18
Holmes, J. H., 43
Hooker, John D., 45, *61*, 70, 98, 193, 194
Hough, George W., 33
House, Edward M., 79
Howard, Robert, 247
Hubble, Edwin, 46, *71*, 196, *236*, 241, *252*
Huggins, Sir William, 16, *54*, 55, 120, 125, 155, 187, 189
Hull, George F., 33
Humason, Milton, 241, *253*
Humboldt, Baron Alexander von, 187
Humphreys, William Jackson, *33*, 76
Huntington, Henry E., 88, 95
Huntington (Henry E.) Library and Art Gallery, 88, *95*, 96
Hussey, William Joseph, 19
Hutchinson, Charles, 19
Huxley, Thomas Henry, 151, 187
Huygens, Christian, 215

Idrac, 76
Ingersoll, Leonard Rose, *50*
Institute of France, 186
International Association of Academies, 177
International Astronomical Union, 73
International Research Council, 73, 80, 82
International Union for Cooperation in Solar Research, 72

Janssen, Pierre Jules César, *16*, 119, 124, 156
Jeans, Sir James, *71*, 286
Jewett, Frank Baldwin, *80*
Johns Hopkins University, 240
Johnson, Harold, 242

Kapteyn, Jacobus Cornelius, 76
Kapteyn, Mrs. J. C., 76
Kathan, George, 33
Kayser, Heinrich Gustav Johannes, 76
Keats, John, 273
Keeler, James Edward, 18, 21, *33*, 211, 223
Kellogg, Vernon Lyman, 85
Kenwood Physical Observatory, 2, 18, 21, *24*, *25*, 51, 98, 239
Kiepenheuer, Karl Otto, 259
King, Arthur Scott, *76*, 170, 239
Kirchhoff, Gustave-Robert, 118, 119
Kitt Peak National Observatory, 212, 215, 216, 218, 241, *256*
Klüber, Harald von, 259
Knight, 76

290   Index

Konen, Heinrich Mathias, *76*
König, Karl Rudolph, 163
Kron, Gerald Edward, 244
Küstner, Friedrich, *76*

La Baume Pluvinel, Eugène-Aymar, *76*
Lallemand, André, 244
Lampland, Carl Otto, *76*
Langer, Robert, 258
Langley, Samuel Pierpont, 17, *41*, 157
Laplace, Pierre Simon, 150
Larkin, Edgar Lucien, *76*
Larmor, Sir Joseph, *76*
Lassell, William, 210, 216, 220
Laves, Kurt, *33*
Leavenworth, Francis Preserved, *33*
Leighton, Robert, 242, 261
Leuschner, Armin Otto, *76*
Lick, James, 19, 98, 194, 209, 217, 218, *222*
Lick Observatory, 18, 19, *23*, 207, 208, 210, 211, 212, 215, 217, *222, 223, 226*, 232
Lick trustees, 209, 210
Lillie, Frank R., 97
Lincoln, Abraham, *272*
Linnean Society, 178
Little, Arthur Dehon, 189
Lockyer, Sir Joseph Norman, 16, 116, 118, 119, 120, 123, 127, 163
Lohse, Ostwald, 122, 123, 125
Loomis, Elias, *176*
Lord, Henry Curwen, *33*
Louville, Chevalier de, 117
Low, Frank, 242
Lundin, Carl A. R., *29, 33*
Lyon, Alva A., *33*

McAdie, Alexander George, *76*
McBride, James H., *61*
McCauley, George V., *105*
McDonald Observatory, 215, *229*
McGee, James Dwyer, 244
McLeod, Clement Henry, *33*
Maddrill, James Davis, *76*
Mantois, 19
Marconi, Guglielmo, 190
Martz, Dowell, 242
Massachusetts Institute of Technology, 2, 9, *10*, 11, 18, 87, 97, 98, 114, *144*, 146, 147, 148, 149, 152, 153, 247
Mayer, Alfred Marshall, *176*
Merriam, John C., 85
Meudon Astrophysical Observatory, 16, *234*
Michelson, Albert Abraham, 45, 85, 87, *176*, 246, 257
Miller, *50*
Miller, Ephraim, *33, 76*
Millikan, Robert Andrews, 87, *92, 254*, 273, *283*
Mitchell, Silas Weir, *176*
Mitchell, Walter Mann, *76*, 163, 164
Moore, Eliakim Hastings, 19
Morgan, Lewis Henry, *176*
Mors, George C., *33*
Mosgrove, Alicia, *70*
Mount Wilson Observatory, 19, 42, 43, 45, *51, 53, 60, 62, 70, 71*, 72, 74, *75, 76*, 77, 87, 88, 97, 99, 100, *112*, 155, 158, 164, 169, 170, *173*, 174, 193, 194, 195, 199, 200, 211, 212, 214, 215, 217, 218, *230, 231*, 238, 239, 240, 241, 243, 245, 246, *251*, 257, 259, 261, *267*, 273, 274
Mount Wilson and Palomar Observatories, 99, *107*, 207, 240, 243; see also Hale Observatories
Muir, John, *61*
Munch, Guido, 246
Myers, George W., *33*

Nasmyth, James, 210
National Academy of Sciences, 72, 73, 80, *84, 85*, 86, 114, 176, 177, 179, 180, 181, 182, 183, 184, 185, 186, 187, 188, 189, 191, 204, 272, 275, 276, 277, 278, 279
Annual Reports, 184
Memoirs, 184
Proceedings, 97, 178, 180, 181, 182, 183, 184, 185, 276
National Research Council, 72, 73, 78, 79, 277, 278, 279, 280, *283*
National Research Fellowships, 73, 280
Nature, 2, 185
Neugebauer, Gerry, 242
Newall, Hugh Frank, 2, *76*, 77
Newberry, John Strong, *176*
Newcomb, Simon, 21, 22, *39, 176*
Newton, Sir Isaac, 195, 198, 199, 273
Nichols, Ernest Fox, *33, 35*, 170, 241
Nicholson, Seth Barnes, 241, 242
Noyes, Arthur A., *10, 40*, 87, *91, 92*
Noyes, Robert Wilson, 261

Oakland Public School, 2
Osborn, Henry Fairfield, *61*

Palomar Observatory, *108, 109, 192*, 209, 211, 212, 214, 217, *231, 236*, 239, 240, 243
Parker, Peter, *176*
Parkhurst, John Adelbert, 21, *33, 35*
Parsons, Charles Algernon, 190
Pasadena Auditorium, 89
Pasadena City Hall, 89
Pasadena and Mount Wilson Toll Road Company, 43
Pasedena Public Library, 89
Pasteur, Louis, 178
Paul, Henry Martyn, *33*
Payne, William Wallace, 21, *33*
Pease, Francis Gladhelm, 45, 64, 196, 200, 201, 246, *255*
Peirce, Benjamin, *176, 272*
Perkin-Elmer Corporation, 240
Perrine, Charles Dillon, 211
Pettit, Edison, 241, 242
Pfeffer, Wilhelm, 179
Pfund, August, 241
Pickering, Edward Charles, 9, 12, *13, 33*, *76, 77*, 127, 138
Pickering, Mrs. E. C., *33*
Plaskett, John Stanley, *76*
Poincaré, Henri, 72
Poor, Charles Lane, *33*
Potsdam Observatory, *27*
Pringsheim, Ernst, *76*
Pritchett, Henry Smith, *33*
Puiseux, Pierre Henri, *76*

Quimby, Rev. Alden Walker, *33*

Ranyard, Arthur Cowper, 42, 199
Rayleigh, Lord (John William Strutt), 185
Rees, J. K., *33*

Rhees, William Jones, *176*
Ricco, Annibale, *76*
Richardson, Robert, 258
Ritchey, George Willis, *33, 35,* 43, *76,* 193, 211, 213
Roberts, Isaac, 210
Robinson, Thomas Romney, 210, 213
Rockefeller, John D., 20, *32*
Rockefeller Foundation, 73, 98, 278, 280
 International Education Board of the, 99
Rockwell, C. H., *33*
Root, Elihu, 278
Rose, Wickliffe, 98, 99, 103
Rosse, Lord (William Parsons), 196, 199, 210, 221
Rotch, Abbott Lawrence, *76*
Rowland, Henry Augustus, 128, 161, 164, 200, *252*
Royal Astronomical Society, 178, 273
Royal Greenwich Observatory, 212, *225*
Royal Institution of Great Britain, 114, 155, 162, 178, 185, 186, 189
Royal Saxon Academy of Sciences, 120
Royal Society of London, 2, 54, 73, 120, 178, 181, 184, 185, 187, 189
 Philosophical Transactions, 178
 Proceedings, 120, 178, 183, 185
Rumford spectroheliograph, 21, *36, 37*
Runge, Carl, *33*
Russell, Henry Norris, *76*
Rutherford, Sir Ernest, 87, 188
Rydberg, Johannes Robert, *76*

Saari, John, 242
St. John, Charles Edward, *76,* 170
Sandage, Allan, 46
Sanford, J. F., *76*
Scherer, James A. B., *61,* 87, *90*
Schlesinger, Frank, *35, 76*
Schmidt, Bernhard, 217, 241, *253*
Schott, Charles Anthony, *176*
Schuster, Arthur, *76,* 170
Schwarzschild, Karl, *76*
Schwarzschild, Martin, 213
Science, 185
Seabroke, George Mitchell, 120, 123
Seares, Frederick Hanley, *33,* 197
Secchi, Angelo, 118, 120, 157
Serrurier, Mark, 215
Shane, C. Donald, *233*
Shorthill, Richard, 242
Sibley College, 147
Sidereal Messenger, 21
Silliman, Benjamin, *176*
Simon, George Warren, 261
Slemp, C. Bascom, 85
Slipher, Vesto Melvin, *76*
Slocum, Frederick, *76*
Smithsonian Institution, 41, *176,* 180
Snyder, Monroe Benjamin, *33*
Spencer, Herbert, 150
Staats, William R., 43
Stannyan, Captain, 117
Stebbins, Joel, 243, 259
Stillhamer, Arthur Grant, *33*
Stratton, Samuel Wesley, *76*
Strong, John, 213, 259
Struve, Otto, 21
Swasey, Ambrose, 18
Swezey, Goodwin Deloss, *33*

Talleyrand-Périgord, Charles Maurice de, 186

Technology Review (M.I.T.), 87, 97
Telescopes
 Catalina Observatory of the Lunar and Planetary Laboratory
  60-inch photometric telescope, 242
  20-inch telescope, 242
 Cerro Tololo Inter-American Observatory
  60-inch telescope, *227*
 Dominion Astrophysical Observatory
  72-inch telescope, 215, *228*
 Galileo's telescope, 194
 Hamburg Observatory
  Schmidt telescope, 217
 Harvard College Observatory
  11-inch telescope, 132
  15-inch telescope, 127, 132
  12-inch telescope, 127, 128, 129, 130, 133, 134, 135, 138, *140, 141, 142*
 Helwan Observatory
  14-inch telescope, 242
 Huggins's 8-inch telescope, *16*
 Kenwood Observatory
  12-inch telescope, 18, 26
 Kitt Peak Observatory
  50-inch telescope, 216
  McMath (Robert M.) solar telescope, 212, *256*
  150-inch telescope, 241
 Lassell's 48-inch telescope, 216, *220*
 Lick Observatory
  120-inch telescope, 207, *208,* 212, 215, *232, 233*
  36-inch Crossley telescope, 210, 211, 213, 215, *223, 226*
  36-inch Lick telescope, *23,* 209, 210, *222*
 McDonald Observatory
  107-inch telescope, *229*
 Mount Wilson Observatory
  100-inch Hooker telescope, 45, 46, 64, 98, 115, 193, 194, 195, 196, 198, 199, 200, 201, 211, 212, 214, 215, *231, 238,* 241, 243, *255,* 260
  150-foot tower solar telescope, 45, 75, 169, 257, 259, 267
  60-foot tower solar telescope, 44, 164, 166, *173*
  60-inch telescope, 42, 43, 45, 60, *64,* 98, 193, 197, 199, 201, 211, 214, 215, *224, 230*
  Snow solar telescope, 42, 44, *52, 112,* 165, 239
 Palomar Observatory
  18-inch Schmidt telescope, 217
  48-inch Schmidt telescope, 217, *236*
  200-inch Hale telescope, 98, 99, 103, *104, 105, 106, 107, 108, 109,* 115, *192,* 196, 202, 207, 209, 211, 212, 214, 215, *224, 231, 235,* 239, 240, 241, 243, 286
 Princeton University Observatory
  6-inch telescope, 133
  36-inch Stratoscope II telescope (balloon), 213, 237
  23-inch telescope, 163
 Roberts's (Isaac) 20-inch telescope, 210
 Rosse's telescope, 199, *221*
 Royal Greenwich Observatory
  98-inch Isaac Newton telescope, 212, 214, *225*
  236-inch telescope (USSR), 211
 U.S. Naval Observatory, Flagstaff
  61-inch telescope, 214

292   Index

Warner and Swasey Observatory
  24-inch Schmidt telescope, 217
Yerkes Observatory
  40-inch telescope, *28, 29, 30,* 31, *36, 42,* 44, 199, 210
  24-inch telescope, 211
Thiessen, Georg Heinrich, 259
Thollon, Louis, 132
Thomson, Elihu, 201
Thomson, Sir Joseph John, *80,* 155, 162, 185
Throop College of Technology, Throop Polytechnic Institute; *see* California Institute of Technology
Townley, Sidney Dean, *76*
Turner, Herbert Hall, *76*
Tyndall, John, 185

United States Naval Observatory, Flagstaff, 214
Universal Exposition of St. Louis, 1904, 72
Updegraff, Milton, *33*
Upton, Winslow, *33*

van der Waals, Johannes Diderick, 179
van't Hoff, Jacobus Hendricus, 190, 191
Van Vleck, John Monroe, *33*
Vassenius, 117
Vaughan, Arthur Harris, 245, 246
Very, Frank Washington, *33*

Wadsworth, Frank Lawton Olcott, *33*
Walcott, Charles Doolittle, 42, 72, *74, 85*
Walcott, Mary Vaux (Mrs. Charles), 42
Walker, 244
Warner and Swasey Company, 18, 20, 215, 217
Washington Academy of Sciences, *Journal,* 184
Watson, *76*
Welch, William H., 73, *80*
Western Electric Company, 80
Whitford, Albert, 243, 259
Whitney, Mary Watson, *33*
Wilson, Henry, *272*
Wilson, Herbert Couper, *76*
Wilson, Woodrow, 73, 78, *79,* 277, 278, 285
 executive order, *284*
Winlock, Joseph, 121
Wolfer, *76*
Woodward, Joseph Janvier, *176*
Woodward, Robert W., 44

Yerkes, Charles Tyson, 19, 20, *29,* 98, 193, 194, 273
Yerkes Observatory, 19, 20, 21, 28, 31, *32, 33,* 34, *35, 38,* 42, 88, 193, 200, 210, 211, 239, 241, *250,* 273
Young, Charles Augustus, 9, *12,* 21, 119, 122, 133, 163

Zeeman, Pieter, 44, *58,* 155, 162, 163, 166
Zenger, Carl Wenzl, 122, 123, 124
Zöllner, Friedrich, 120, 121